Lecture Notes in Networks and Systems

Volume 52

Series editor

Janusz Kacprzyk, Polish Academy of Sciences, Warsaw, Poland
e-mail: kacprzyk@ibspan.waw.pl

The series "Lecture Notes in Networks and Systems" publishes the latest developments in Networks and Systems—quickly, informally and with high quality. Original research reported in proceedings and post-proceedings represents the core of LNNS.

Volumes published in LNNS embrace all aspects and subfields of, as well as new challenges in, Networks and Systems.

The series contains proceedings and edited volumes in systems and networks, spanning the areas of Cyber-Physical Systems, Autonomous Systems, Sensor Networks, Control Systems, Energy Systems, Automotive Systems, Biological Systems, Vehicular Networking and Connected Vehicles, Aerospace Systems, Automation, Manufacturing, Smart Grids, Nonlinear Systems, Power Systems, Robotics, Social Systems, Economic Systems and other. Of particular value to both the contributors and the readership are the short publication timeframe and the world-wide distribution and exposure which enable both a wide and rapid dissemination of research output.

The series covers the theory, applications, and perspectives on the state of the art and future developments relevant to systems and networks, decision making, control, complex processes and related areas, as embedded in the fields of interdisciplinary and applied sciences, engineering, computer science, physics, economics, social, and life sciences, as well as the paradigms and methodologies behind them.

Advisory Board

More information about this series at http://www.springer.com/series/15179

Elżbieta Macioszek · Rahmi Akçelik
Grzegorz Sierpiński

Editors

Roundabouts as Safe and Modern Solutions in Transport Networks and Systems

15th Scientific and Technical Conference
"Transport Systems. Theory and Practice 2018",
Katowice, Poland, September 17–19, 2018,
Selected Papers

 Springer

Editors
Elżbieta Macioszek
Faculty of Transport
Silesian University of Technology
Katowice, Poland

Grzegorz Sierpiński
Faculty of Transport
Silesian University of Technology
Katowice, Poland

Rahmi Akçelik
Akcelik and Associates Pty Ltd
Balwyn North, VIC, Australia

ISSN 2367-3370   ISSN 2367-3389 (electronic)
Lecture Notes in Networks and Systems
ISBN 978-3-319-98617-3   ISBN 978-3-319-98618-0 (eBook)
https://doi.org/10.1007/978-3-319-98618-0

Library of Congress Control Number: 2018950541

This Springer imprint is published by the registered company Springer Nature Switzerland AG
The registered company address is: Gewerbestrasse 11, 6330 Cham, Switzerland

Preface

Roundabouts are characterized by many benefits, which primarily include a high level of road safety, significant efficiency and smoothness of traffic. These benefits make roundabouts very popular in many countries around the world. However, these intersections are not necessarily the right solution for all traffic conditions and for all locations in a transport network. In the case of high demand volumes at this type of intersection, the vehicle delays increase rapidly. A basic condition for a correctly designed intersection is to ensure an adequate level of service for its users. In order to be able to make assessments related to this, designers, planners and other people dealing with transport must have at their disposal methods for estimating road traffic conditions including capacity and delay at a given intersection. Tools for assessment of pollutant emissions, planning of roadworks, application of temporary traffic control, evaluation of vehicle traffic trajectory and assessment of applied solutions for unprotected users of roundabouts are also needed.

In addition, contemporary technical and technological developments, dynamics of changes in the modern world, complexity of road infrastructure as well as the ever-growing mobility of contemporary society provide favourable conditions for designing innovative transport-related solutions, including various new solutions for roundabouts.

The book, entitled *Roundabouts as safe and modern solutions in transport networks and systems*, provides an excellent opportunity to become familiar with the latest solutions for roundabouts as particularly desirable elements of transport networks and systems both in Poland and other countries of the world. The presentations allow familiarity with the challenges and innovative directions of research and analysis related to roundabouts. The book has been divided into four parts. These are:

- Part 1. Modern Solutions for Roundabouts—the Latest State of the Art,
- Part 2. Roundabouts as Elements of Smart and Modern Transport Networks and Systems,

- Part 3. Data Collection, Data Analysis and Development of Model and Methods for Roundabouts,
- Part 4. Road Traffic Safety Analysis.

The publication contains selected papers submitted to and presented at the 15th Scientific and Technical Conference "Transport Systems. Theory and Practice", organized by the Department of Transport Systems and Traffic Engineering at the Faculty of Transport of the Silesian University of Technology (Katowice, Poland). The papers cover current modelling issues and methods used in the assessment of the efficiency of roundabouts. They also deal with data collection and data analysis methods. Numerous practical examples present various novel solutions and indicate the latest state of the art in research and development related to roundabouts. While roundabouts are considered to exert significant influence on increasing the functional efficiency of transport networks and systems, they provide priority for the health of people, road traffic safety, sustainable development of transport systems and protection of the natural environment. Accordingly, a part of the book relates to the problems associated with road traffic safety analysis.

We would like to use this occasion to express our gratitude to the authors for submitting their papers to share the results of their scientific and research work. Their contributions illustrate how safe and modern solutions like roundabouts play an important role in transport networks and systems in view of the multiple challenges of the contemporary world facing them. We would also like to thank the reviewers from many countries around the world, sometimes very distant, for providing a global look at the issues covered by these papers. We thank them for their insightful remarks and suggestions which have ensured the high quality of this publication.

Readers interested in roundabout issues will find in this volume a wide range of research material presenting the results of scientific research, diverse insights and comments as well as new approaches and solutions to problems. With this in mind, we hope that readers find this book valuable.

September 2018

Elżbieta Macioszek
Rahmi Akçelik
Grzegorz Sierpiński

Organization

15th Scientific and Technical Conference "Transport Systems. Theory and Practice" (TSTP 2018) is organized by the Department of Transport Systems and Traffic Engineering, Faculty of Transport, Silesian University of Technology (Poland).

Organizing Committee

Organizing Chair

Grzegorz Sierpiński   Silesian University of Technology, Poland

Members

Renata Żochowska
Grzegorz Karoń
Krzysztof Krawiec
Aleksander Sobota
Marcin Staniek
Ireneusz Celiński

Barbara Borówka
Kazimierz Dąbała
Marcin J. Kłos
Damian Lach
Piotr Soczówka

The Conference Took Place Under the Honorary Patronage

Marshal of the Silesian Voivodeship

Scientific Committee

Stanisław Krawiec (Chairman)	Silesian University of Technology, Poland
Rahmi Akçelik	Sidra Solutions, Australia
Tomasz Ambroziak	Warsaw University of Technology, Poland
Henryk Bałuch	The Railway Institute, Poland
Roman Bańczyk	Voivodeship Centre of Road Traffic in Katowice, Poland
Werner Brilon	Ruhr University Bochum, Germany
Margarida Coelho	University of Aveiro, Portugal
Boris Davydov	Far Eastern State Transport University, Khabarovsk, Russia
Mehmet Dikmen	Baskent University, Turkey
Domokos Esztergár-Kiss	Budapest University of Technology and Economics, Hungary
József Gál	University of Szeged, Hungary
Andrzej S. Grzelakowski	Gdynia Maritime University, Poland
Mehmet Serdar Güzel	Ankara University, Turkey
Józef Hansel	AGH University of Science and Technology, Cracow, Poland
Libor Ižvolt	University of Žilina, Slovakia
Marianna Jacyna	Warsaw University of Technology, Poland
Nan Kang	Tokyo University of Science, Japan
Jan Kempa	University of Technology and Life Sciences in Bydgoszcz, Poland
Michael Koniordos	Pireaus University of Applied Sciences, Greece
Bogusław Łazarz	Silesian University of Technology, Poland
Zbigniew Łukasik	Kazimierz Pulaski University of Technology and Humanities in Radom, Poland
Michal Maciejewski	Technical University of Berlin, Germany
Elżbieta Macioszek	Silesian University of Technology, Poland
Ján Mandula	Technical University of Košice, Slovakia
Sylwester Markusik	Silesian University of Technology, Poland
Antonio Masegosa	IKERBASQUE Research Fellow at University of Deusto, Bilbao, Spain
Agnieszka Merkisz-Guranowska	Poznań University of Technology, Poland
Anna Mężyk	Kazimierz Pulaski University of Technology and Humanities in Radom, Poland
Maria Michałowska	University of Economics in Katowice, Poland
Leszek Mindur	International University of Logistic and Transport in Wrocław, Poland
Maciej Mindur	Lublin University of Technology, Poland

Referees

Rahmi Akçelik
Marek Bauer
Przemysław Borkowski
Werner Brilon
Margarida Coelho
Piotr Czech
Domokos Esztergár-Kiss
Michal Fabian
Barbara Galińska
Róbert Grega
Mehmet Serdar Güzel
Katarzyna Hebel
Nan Kang
Peter Kaššay
Jozef Kuĺka
Michał Maciejewski
Elżbieta Macioszek
Krzysztof Małecki

Martin Mantič
Silvia Medvecká-Beňová
Katarzyna Nosal Hoy
Romanika Okraszewska
Asier Perallos
Hrvoje Pilko
Antonio Pratelli
Michal Puškár
Piotr Rosik
Alžbeta Sapietová
Grzegorz Sierpiński
Marcin Staniek
Dariusz Tłoczyński
Andrzej Więckowski
David Williams
Grzegorz Wojnar
Adam Wolski
Ninoslav Zuber

Contents

Modern Solutions for Roundabouts - the Latest State of the Art

Turbo Roundabouts: A Brief Safety, Efficiency and Geometry Design Review

Hrvoje Pilko[(✉)], Željko Šarić, and Goran Zovak

Faculty of Transport and Traffic Sciences, University of Zagreb, Zagreb, Croatia
{hpilko,zsaric,gzovak}@fpz.hr

Abstract. The application popularity of various roundabout types around the world has driven substantial efforts to optimize their geometry design. Implementing effective roundabouts requires optimizing traffic safety (TS) and traffic (operational) efficiency (TE) while considering various geometry and other factors. In multi-lane roundabouts driver indecision and misunderstanding of the driving rules and situations can lead to weaving conflicts and accidents. These accidents are frequent and often affect TS and TE. The turbo roundabout concept has emerged as a possible alternative to conventional multi-lane roundabouts, aiming to improve exclusively TS. However, some studies do not allow definitive conclusions about their TE (capacity) influence. To capture a relationship between turbo roundabout design, TS and TE parameters brief overview of the latest guidelines and studies is presented here.

Keywords: Turbo roundabout · Traffic safety · Traffic (operational) efficiency
Guidelines · Geometry design

1 Introduction

The popularity of roundabouts around the world has driven substantial efforts to optimize their planning and design procedures. Implementing roundabouts is a highly demanding task and requires optimizing traffic safety (TS) and traffic (operational) efficiency (TE) while considering geometry factors, traffic characteristics and local constraints. Studies of roundabouts in various countries, particularly of single-lane roundabouts in urban areas, have shown that proper design can significantly improve TE, as well as TS parameters [1, 2]. In multi-lane roundabouts driver indecision and misunderstanding of the driving rules and situations can lead to weaving conflicts and accidents. These accidents, although not usually severe, are frequent and often affect TE.

The turbo roundabout concept has emerged as a possible alternative to conventional multi-lane roundabouts, aiming to improve TS. However, studies do not allow definitive conclusions about their TE and thus further research is recommended [3]. The conventional multi-lane roundabout is an efficient solution to cope with higher traffic flow demand. Additional entry and circulatory lanes increase TE parameters but they also increase the TS. These are mainly related to improper driver behavior at the entrance, circulatory and exit zones, and to the consequent weaving maneuvers within

E. Macioszek et al. (Eds.): Roundabouts as Safe and Modern Solutions in Transport
Networks and Systems, LNNS 52, pp. 3–12, 2019.
https://doi.org/10.1007/978-3-319-98618-0_1

the circle. At multi-lane roundabouts, increasing vehicle path curvature creates greater side friction between adjacent traffic streams and can result in more vehicles cutting across lanes and higher potential for sideswipe collisions [3].

Previous studies on two-lane roundabouts confirmed improper behavior as being common practice, resulting in conflicts and increased likelihood of crashes [4]. Also, the use of the inner circulatory lane is poor, which will have a negative impact on the capacity [5]. About 40% of drivers in Portugal who approach the roundabout using the right lane invade the left circulatory lane, thus following a rectilinear path. About 20% of the drivers who approach the roundabout using the left lane take the right exit lane, to minimize driving inconvenience. Some countries have avoided these problems by limiting the adoption of multi-lane roundabouts; France, Germany and Switzerland [4]. Despite this attempt to increase TS, accidents at roundabouts are common so turbo roundabout concept has emerged [4].

The focus of the research is to study the impact of turbo roundabouts as a new roundabout concept design. This is done by using the latest relevant literature review considering TS and TE parameters, design guidelines and geometry elements.

The remainder of the paper is organized as follows. Section 2 briefly describes the turbo roundabout design concept: latest guidelines, geometry design with Croatian experience. Section 3 briefly presents TS and Sect. 4 TE latest findings. Section 5 discusses the results and implications for future work.

2 Design Principles

2.1 Guidelines

Previous turbo roundabouts studies were mainly based on evaluation of their TE (capacity and delays), and environmental, economic and TS parameters, regarding single-lane, multi-lane, flower and target roundabouts [6]. First turbo roundabout guidelines were published in the Netherlands in 2008 [7]. These guidelines were widely accepted and used by road designers in practice. Slovenian government released a draft version of Slovenian technical specifications on turbo roundabouts in 2011 [8].

In 2015 German association FGSV issued an official working document on turbo roundabout use and design [9]. Serbian Authority for Roads published a turbo roundabout guideline [10] in 2014. Unlike the other countries with technical regulations on turbo roundabouts, Serbia has not build any turbo roundabouts yet. Croatian turbo roundabout guidelines [11] were published in 2014, when the first turbo roundabout in Croatia was built (Fig. 3). Nowadays there are few of them located in the city of Osijek and Pula.

Comparative analysis of turbo roundabout design procedures described in Dutch, Slovenian, German, Serbian and Croatian guidelines on turbo roundabouts is in detail given in [6], with brief overview here. Dutch, Slovenian and German guidelines are chosen for the analysis since these guidelines origin from the countries with a notable experience in turbo roundabout design. Serbian and Croatian guidelines are chosen because they are the most recent turbo roundabout guidelines. According to Slovenian,

Serbian, German, Croatian and Dutch guidelines, geometry design of turbo roundabouts can be carried out through the following steps: (1) selecting one of the available roundabout types; (2) defining a relevant design vehicle; (3) creating one of given turbo block templates; (4) designing the remaining turbo roundabout elements; and (5) conducting design vehicle horizontal swept path analysis and fastest path vehicle speed analysis.

Major guideline differences are shown in the following. Main roundabout forms given in Dutch, Slovenian and Serbian guidelines are: four leg variants that could also be planned as three leg variants: Egg, Basic turbo, Knee, Spiral and Rotor roundabout; and three leg variants: Stretched-knee and Star roundabout (Fig. 1). Egg, Basic turbo, Knee, Spiral and Stretched-knee roundabout are recommended forms when one of the traffic demand flows is predominant. Rotor or Star roundabout forms are recommended in case of equal traffic demand flows on all approaches. Other modified variants can further be designed by varying the number of entry lanes. Croatian guidelines define roundabout forms: Egg, Basic turbo, Knee and Stretched-Knee roundabout. According to German guidelines, there are maximum two traffic lanes at roundabout circulatory roadway, and several possible arrangements of entry and exit lanes on roundabout approaches.

2.2 Geometric Elements

According to observed guidelines Džambas [6] concluded that turbo roundabout elements must provide unobstructed passage of the relevant design vehicle. Design vehicle and its swept path trajectories have a strong influence on the turbo roundabout geometry design. The choice of the relevant design vehicle should be based on the structure of traffic demand flow, which significantly depends on the share of the vehicle in the vehicle fleet of the region where the roundabout is located and local regulations. Swept path trajectories of relevant design vehicle must be checked using one of the CAD program tools, usually AutoCAD with AutoTurn, where all relevant design vehicles can be found defined by guidelines.

Croatian guidelines provide dimensions for various turbo roundabout templates, but it is unclear to which design vehicle they are associated with. German guidelines states that all turbo roundabout elements must be designed regarding design vehicle swept path. Despite this, relevant design vehicle for turbo roundabout design is not recommended. Slovenian and Serbian guidelines provide that relevant design vehicle is usually a 16.50 m long truck with a semitrailer.

Design vehicle parameters that influence vehicle swept path width are: overall width w, length of the front overhang t, and length of the wheelbase s_2. Design vehicles have similar lengths of the front overhang and the wheelbase, but different overall widths: Croatian, German, Serbian and Slovenian design vehicles are 2.50 m wide, and Dutch 2.55 m. Because wider vehicles occupy a greater area when driving critical turning movement, it would be advisable that widths of the design vehicles are set to 2.55 m [6].

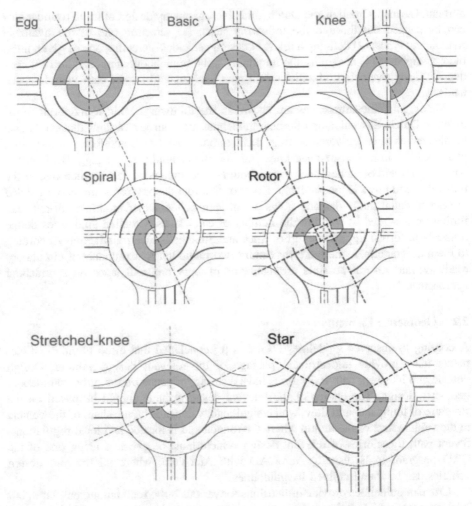

Fig. 1. Turbo roundabout variants with four and three approaches. (Source: [6])

Turbo roundabout geometry design is mainly defined by defining the turbo block. There can be two groups: guidelines that provide turbo block templates with predetermined dimensions [7, 8, 10, 11], and guidelines that do not provide turbo block templates [9]. Recommended maximum width of circulatory lane is 5.25 m. Turbo block should be designed in a way that circular arcs at one side of the translation axis overlap with circular arcs at the other side of the translation axis i.e. that inner circular lane at one side of the translation axis continues the outer circular lane at the other side of translation axis [3]. Turbo block templates given in Dutch, Slovenian and Serbian guidelines do not entirely fulfill this requirement. In these templates 5 cm shift of circular arcs at translation axis exists (Fig. 2). In Croatian guidelines this shift is eliminated by application of 5 cm wider outer marginal strips i.e. circular arcs are overlapping on translation axis.

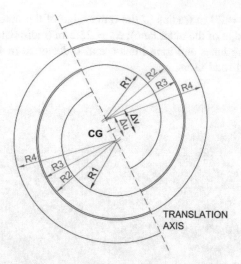

Fig. 2. Turbo block elements for common Dutch, Croatian, Slovenian and Serbian roundabout variants. (Source: [6])

After designing a turbo block, remaining turbo roundabout elements can be designed: central island, approaches, and raised mountable lane dividers. Detail guideline comparison can be found in [6].

2.3 Croatian Example

The procedures of planning and designing the roundabouts in the Republic of Croatia are based on the current national and applied foreign guidelines, especially Netherland, German, Austrian and Switzerland, positive examples of world practice and empirical practice of designers. One of the first steps in creating national regulations for roundabout designing were the Guidelines for the Design of Circular Intersection (Guidelines2002) from 2002 [12]. The main objective of these guidelines was the standardization of design and implementation of roundabouts on public roads in the country. Also, insufficient attention has been made to geometry design and the concept of turbo roundabouts [1].

Roundabout Guidelines for State Roads (Guidelines2014) from 2014 [11] represent a significant upgrade of Guidelines2002 in terms of geometry design elements, TE criteria, defining, planning and implementation of turbo roundabouts, and the importance and necessity for swept path trajectory checking of the relevant design vehicle in the design phase. However, the guidelines do not indicate what models or simulation software should be used for analyzing TE.

The first turbo roundabout in Croatia was carried out in the city of Osijek in 2014 (Fig. 3). Intersection is designed in accordance with the Guidelines2014. The most important design elements are: external diameter is 5.15 m (outer space between shifted centers of circular segments); inscribed central diameter is 4.95 m (inner space between shifted centers of circular segments); R_1 = 15 m (radius of the inner edge of the road surface); $R_2 = 20$ m (radius of the outer edge of the outer lane); $R_3 = 20.3$ m (radius of the inner edge of the outer lane); $R_4 = 25.2$ m (radius of the outer edge of the road surface). Cycling lanes and tram lines are also designed making this intersection more complex for all road users.

Fig. 3. Turbo roundabout in the City of Osijek, Croatia - in traffic 2014

Studies on roundabouts in Croatia are on TS and TE parameters, main geometry design elements, vehicle design speeds through the roundabout and observed vehicle speeds [1]. Lately Džambas [6] has studied turbo roundabouts and its performance.

3 Traffic Safety

Turbo roundabouts have two principal advantages over conventional double-lane roundabouts, based on the physical separation of the lanes [4]: (1) reduction in the number of conflict points, and (2) speed reduction along the entry, circulatory and exit zones. Other studies, based on conflict analysis techniques applied to nine layouts with different demand scenarios, show a 40–50% reduction in the accident rate [4]. In a study based on microsimulation applications, [4] concluded that drivers using the outer lane of a turbo roundabout drive more slowly than in the double-lane roundabout, with reductions from 48 to 38 km/h. There is a reduction from 24 conflict points on the double-lane roundabout to 14 points on the turbo-roundabout, indicating an overall reduction in crash probability. It should be noted, however, that some of these conflicts exhibit higher severity both because of the increased impact angle and because circulating traffic is concentrated in the outer lane. In the absence of historic crash data, a deeper analysis is needed using microsimulation techniques (i.e. PTV VISSIM).

4 Traffic Efficiency

As the TS positive effects are recognized, there are still doubts about TE parameters, concerning capacity. Researchers have been using methods that not fully describe the complex interactions between the different traffic flows. Fixed lane usage at the entries and the irrelevance of traffic distribution in the circulatory lanes are fully described in a new calculation method [4] based on gap-acceptance theory and on the generalization of Tanner's formula for multiple traffic lanes, briefly described here. The capacity of a single-entry lane that merges or crosses a single circulating line is than given by [4]:

$$C = \frac{q_c \phi e^{-\lambda(t_c - \Delta)}}{1 - e^{-\lambda t_f}} \tag{1}$$

where:

Q - capacity of the entry [veh/s],
q_c - the conflicting flow [veh/s],
t_c - the critical headway [s],
t_f - the follow-up time [s].
$\emptyset, \lambda, \Delta$ - parameters of the Cowan M3 distribution.

Bastos [4] conclude that, for Portuguese conditions and assuming a fixed intra-platoon headway $\Delta = 2$ s, the parameters $\emptyset$ and λ can be calculated using Eqs. (2) and (3):

$$\phi = \begin{cases} 1 & if \quad < 0.178 \\ 1.553(1 - 2q_c) & if \quad 0.178 < q_c \le 0.5 \\ 0 & if \quad q_c > 0.5 \end{cases} \tag{2}$$

$$\lambda = \frac{\phi q_c}{1 - \Delta q_c} \tag{3}$$

For entries with two circulating lanes the capacity is given by Eq. (4). The indexes 1 and 2 stand for each of the opposing lanes in the circulatory roadway (1 - outer lane, nearest the entry; 2 - inner lane, near the circle). The parameters $\emptyset$ and λ should be calculated using the Eqs. (2) and (3), with $\Delta = 2$ s. This equation is used separately for each entry lane. The parameters t_c and t_f can take different values according to the entry lane (left/right) [4]:

$$C = \frac{\exp[-(\lambda_1 + \lambda_2)(t_c - \Delta)](\lambda_1 + \lambda_2)\phi_1\phi_2}{\langle 1 - \exp[-t_f(\lambda_1 + \lambda_2)]\rangle(\phi_1 + \lambda_1\Delta)(\phi_2 + \lambda_2\Delta)} \tag{4}$$

A sensitivity analysis of Eq. (4) indicates that the maximum capacity is achieved when the opposing traffic is evenly distributed in the circulatory lanes. In fact, this split increases the probability of side-by-side or overlapped circulation, thus reducing the waste of opportunities for the waiting vehicles. Considering that there are no U-turn maneuvers, in an equilibrium state the lanes have the same level of saturation and the proportion of through traffic using the inner lane is given by [4]:

$$p_1 = \frac{C_1(q_2 + q_3) - C_o q_1}{q_2(C_1 + C_0)}, p_1 \in [0, 1] \tag{5}$$

where:

C_1, C_O - the inner (left) and outside (right) lane capacities,

$q1$, $q2$, $q3$ - the demand flows for the left, through and right movements respectively.

Similarly, the proportion of right-turning traffic using the inside lane at a minor entry of a turbo roundabout [4]:

$$p_1 = \frac{C_1 q_3 - C_0(q_1 + q_2)}{q_3(C_1 + C_0)}, p_1 \in [0, 1] \tag{6}$$

The cited study reveals that the rigid allocation of movements to the entry lanes in the turbo roundabout's minor often leads to higher saturation in the left lane and consequently to a waste of the right lane's capacity (used only for the right turns). This does not happen in the major direction because drivers can select the less congested entry for the through movements. It can be concluded that turbo roundabouts offer more capacity than two-lane roundabouts of similar size only in specific and rare circumstances of traffic demand, i.e. when the proportion of right turns in the minor direction is very high >60% [4].

5   Discussion and Conclusion

Regarding turbo roundabout design guidelines presented here, we can conclude that they differ in the following: number of turbo roundabout variants, choice of relevant design vehicle(s), dimensions of certain turbo block elements and definition of specific roundabout elements [13]. Nevertheless, turbo roundabout planning procedures are quite similar. In Slovenian, Serbian, Dutch and Croatian guidelines firstly initial roundabout scheme is designed, and then swept path and fastest path vehicle design speed analyses are carried out. This design approach therefore greatly depends on the quality of performance checks, and gives freedom to the designer on the decision whether the project solution is acceptable or not. Studies performed at the Department for Transportation of the Faculty of Civil Engineering, University of Zagreb [14], have confirmed that this design approach ensures the usage of optimal roundabout element dimensions and an unhindered path for the design vehicle through the intersection.

The turbo roundabout concept solution emerged as a way to solve the TS problems of multi-lane roundabouts and for now it succeded. The geometry design elements of these roundabout types effectively impose minimum deflection levels and vehicle design speed, reduce conflict points leading to safer driver and traffic flow operations. These conclusions are consistent with international experience. In terms of TE (capacity), the results are not consensual. Some authors concluded that turbo roundabouts offer better capacity than conventional roundabouts of similar size when the proportion of right turns in the minor direction is very high >60%. The application of a new lane-based method reveals that only in very specific scenarios that are uncommon in real-world networks can a standard turbo roundabout be expected to provide more capacity than the equivalent two-lane roundabout [4].

Future study should involve a greater number of other European and Non-European guidelines and published scientific manuscripts to compare the variety of design concept and usage. Also, to conclude which turbo roundabout types and in what situations/locations can perform better in terms of all TE parameters and it's signalization [15].

Acknowledgement. The research described in this paper was conducted within the scope of the research project Technical Review of Traffic Accident Locations with Killed Persons in the Function of Hazard Identification, funded by the Ministry of the Interior, Republic of Croatia in the scope of the National Road Safety Program 2011–2020.

References

1. Pilko, H., Mandžuka, S., Barić, D.: Urban single-lane roundabouts: a new analytical approach using multi-criteria and simultaneous multi-objective optimization of geometry design, efficiency and safety. Transp. Res. Part C Emerg. Technol. **80**, 257–271 (2017)
2. Hatami, H., Aghayan, I.: Traffic efficiency evaluation of elliptical roundabout compared with modern and turbo roundabouts considering traffic signal control. Promet **29**(1), 1–11 (2017)
3. Fernandes, P., Pereira, S.R., Bandeira, J.M., Vasconcelos, L., Silva, A.B., Coelho, M.C.: Driving around turbo-roundabouts vs. conventional roundabouts: are there advantages regarding pollutant emissions? Int. J. Sustain. Transp. **10**(9), 847–860 (2016)
4. Silva, A.B., Vasconcelos, L., Santos, S.: Moving from conventional roundabouts to turbo-roundabouts. Procedia - Soc. Behav. Sci. **111**, 137–146 (2014)
5. Fortuijn, L.G.H.: Turbo roundabouts. Transp. Res. Rec. **2130**, 83–92 (2010)
6. Džambas, T., Ahac, S., Dragčević, V.: Geometric design of turbo roundabouts. Teh. Vjesn. Tech. Gaz. **24**, 309–318 (2017)
7. CROW: Turborotondes, Publication No. 257. CROW, Netherland (2008)
8. Ministarstvo Za Infrastrukturo in Prostor DRS Za Ceste: Krožna Križišča - TSC 03.341: 2011. Ministarstvo Za Infrastrukturo in Prostor DRS Za Ceste, Ljubljana (2011)
9. FGSV: Handbuch für die Bemessung von Strassenverkehrsanlagen. FGSV, Köln (2015)
10. Serbian Authority for Roads: Priručnik za Projektovanje Puteva u Republici Srbiji, Dio 5.3 Kružne Raskrsnice. Serbian Authority for Roads, Beograd (2012)

11. Deluka-Tibljaš, A., Tollazzi, T., Barišić, I., Babić, S., Šurdonja, S., Renčelj, M., Pranjić, I.: Smjernice Za Projektiranje Kružnih Raskrižja Na Državnim Cestama. Hrvatske Ceste, Zagreb, Rijeka (2014)
12. Institute of Transport and Communications: Smjernice Za Projektiranje i Opremanje Raskrižja Kružnog Oblika - Rotora. Institute of Transport and Communications, Zagreb (2002)
13. Fortuijn, L.G.H.: Turbo roundabouts: design principles and safety performance. J. Transp. Res. Board **2096**, 16–24 (2009)
14. Džambas, T., Ahac, S., Dragčević, V.: Design of turbo roundabouts based on the rules of vehicle movement geometry. J. Transp. Eng. **143**(7), 7–10 (2016)
15. Xu, H., Zhang, K., Zhang, D.: Multi-level traffic control at large four-leg roundabouts. J. Adv. Transp. **50**, 988–1007 (2016)

Work Zones and Temporary Traffic Organization at Roundabout - Review of Selected Solutions

Piotr Soczówka[✉] and Renata Żochowska

Faculty of Transport, Silesian University of Technology, Katowice, Poland
{piotr.soczowka, renata.zochowska}@polsl.pl

Abstract. In many countries roundabouts are becoming an increasingly common element of transportation infrastructure. Therefore, like other elements of the infrastructure, they may be subject to periodic closures associated with i.e. modernization of infrastructure, maintenance of road signs and traffic safety devices or renovation of road surface. Such closures require implementation of temporary traffic organization. Due to unique features of roundabouts it appears that specific methods of organizing work zones at such objects should be developed. This paper contains a proposal of typology of temporary traffic organization at roundabouts and a review of possible solutions for each group.

Keywords: Roundabout · Temporary traffic organization · Disruption
Work zone

1 Introduction

Roundabout is one of the most common type of the intersection. It is generally used due to its high safety level [1] and traffic-calming properties [2]. Roundabouts are considered to be safer than traditional intersections without traffic lights because of lower crash frequencies [3, 4] caused mainly by a smaller number of conflict points [2, 5]. Furthermore, usually the severity of traffic incidents at roundabouts is lower than at other intersections [4]. At the roundabout traffic flows around a central island, which obligates drivers to reduce their speed. Also, in most cases drivers who enter the roundabout must give way to those, who are on the circulatory roadway, which not only forces them to decrease their speed, but also provides opportunity to move more smoothly [6, 7]. Other reasons for selection of roundabout may include i.e. shorter vehicular delays [6, 8] in comparison to other types of intersection.

Although the number of roundabouts has been increasing in recent years (i.e. in United States, total number of roundabouts changed from less than 500 in 2000 to more than 3000 in 2013 [8]) there are only a few studies on temporary traffic organization and work zones at such objects. It seems to be of especially high importance, given that recently constructed roundabouts will have to be maintained or modernized in future. Therefore, it is necessary to work out temporary traffic organization solutions which should be favored, depending on which part of a roundabout is closed.

E. Macioszek et al. (Eds.): Roundabouts as Safe and Modern Solutions in Transport
Networks and Systems, LNNS 52, pp. 13–23, 2019.
https://doi.org/10.1007/978-3-319-98618-0_2

Temporary traffic control is implemented when a planned or not planned event happens, that causes disruptions in traffic flow. Unplanned events (like traffic incidents or sudden failure of traffic control devices) usually require quick actions in order to minimalize negative influence on traffic, and it is difficult to prepare for such events. However, it is still possible to identify most hazardous road locations [9] and prepare plans of temporary traffic organization that may be quickly implemented if needed. On the other hand, planned temporary traffic control may be implemented in a manner that allows to minimalize disruptions caused by different events.

There are various reasons for introducing planned temporary traffic organization at roundabouts, however they in general may be divided into two separate groups [9]:

- reasons connected with maintenance and functioning of traffic infrastructure and traffic devices, such as:
  - maintenance of vertical and horizontal traffic signs,
  - maintenance of road surface,
  - maintenance of traffic safety devices,
  - maintenance of system of dewatering of the road,
- reasons which are not connected with management of road or traffic, such as:
  - performing works on the road,
  - placing devices not connected with traffic management or road management purposes,
  - placing advertisements,

In addition, the introduction of the planned temporary traffic organization may be related to:

- reconstructing the roundabout (i.e. changing the number of lanes at a given element of a roundabout) [8],
- changing the type of a roundabout (i.e. from mini-roundabout to small roundabout).

Although temporary traffic control is to minimalize the negative influence of disruptions traffic condition may become worse. Thus, in a process of planning temporary traffic control it is of great importance to provide as smooth traffic as possible, and as high level of safety as possible [9]. It is essential to provide safe conditions not only for drivers which enter temporary traffic area, but also for workers in a work zone [10, 11].

Because of unique features of roundabout (i.e. circular movement of vehicles around a central island) it is necessary to approach to temporary traffic organization at this element of transportation network in a manner that allows to take advantage of its features.

The purposes of the paper include formulation of typology of possible temporary traffic organization on a roundabout, depending on location of work zone and available width of element of a roundabout and also a review of selected cases of temporary traffic organization on roundabouts.

2 Typology of Possible Traffic Organization on a Roundabout

All works at roundabouts may be performed simultaneously in one or more of following three locations:

- entrance to the roundabout,
- exit from the roundabout,
- circulatory area of the roundabout.

Table 1. Detailed typology of cases of temporary traffic organization.

Location of work zone	Number of traffic lanes at an element of a roundabout in typical conditions	GROUP I	GROUP II	GROUP III
Entrance	$k = 1$	Case I.1	Case II.1	–
	$k > 1$	Case I.2	Case II.2	Case III.1
Exit	$k = 1$	Case I.3	Case II.3	–
	$k > 1$	Case I.4	Case II.4	Case III.2
Circulatory area	$k = 1$	Case I.5	Case II.5	–
	$k > 1$	Case I.6	Case II.6	Case III.3
Entrance and circulatory area	$k = 1$	Case I.7	Case II.7	–
	$k > 1$	Case I.8	Case II.8	Case III.4
Exit and circulatory area	$k = 1$	Case I.9	Case II.9	–
	$k > 1$	Case I.10	Case II.10	Case III.5
Entrance, exit and circulatory area	$k = 1$	Case I.11	Case II.11	–
	$k > 1$	Case I.12	Case II.12	Case III.6
Entrance and exit	$k = 1$	Case I.13	Case II.13	–
	$k > 1$	Case I.14	Case II.14	Case III.7

Temporary traffic organization solution depends on number of traffic lanes which are available for traffic, after full or partial closure of an element of a roundabout. Traffic lane is considered to be available, when it may be used by one stream of vehicles. A set K_{av} of available traffic lanes has been determined as follows:

$$K_{av} = \{k_{av} : (k_{av} \in N \cup \{0\}) \wedge (k_{av} \leq k)\}, \qquad (1)$$

where:

k_{av} - number of available traffic lanes at an element of a roundabout during temporary traffic organization,

k - number of traffic lanes at an element of a roundabout in typical conditions.

There are three possible categories of availability of traffic lanes:

- GROUP I: ($k_{av} = k$) - when the same number of streams of vehicles may go through a given element of a roundabout,
- GROUP II: ($k_{av} = 0$) - when a given element of a roundabout is closed for traffic and no vehicles may use it,
- GROUP III: ($k_{av} \in K_{av}\backslash\{0, k\}$) - when an element of a roundabout is partly closed - it still may be open for traffic, but the number of streams of vehicles which may use it is smaller.

The summary of possible cases of temporary traffic organization is presented in Table 1.

Usually, cases assigned to group 1 cause small difficulties for road users. Difficulties associated with group 3 may be assessed as average whereas group 2 contains cases with large level of difficulties.

3   Overview of Possible Temporary Traffic Control Solutions for Roundabouts

3.1   Cases Assigned to GROUP 1

When temporary traffic organization is assigned to GROUP I, then work zone at the roundabout does not affect traffic in a significant way. The same number of traffic lanes is available for drivers as in typical conditions, however their width may be changed. Nevertheless, it is of great importance to provide roundabout users with all necessary information.

In order to retain a traffic lane in use a minimum unobstructed width is required. English practice [12] suggests that the minimum width should be between 3.25–3.50 m with 3.0 m being an absolute minimum (or 2.5 m when only cars and light vehicles are allowed at a roundabout). According to Polish rules, a minimum width of a single traffic lane is 3.0 m [13].

Work zone should be appropriately separated from traffic for safety of road users and workers as well. If safe separation of the work zone causes that the minimum unobstructed width of a traffic lane cannot be provided the lane should be considered as not available and temporary traffic organization should be assigned to GROUP II or III (for multi-lane elements of a roundabout).

Each roundabout should be treated individually when organizing safe work zone. Different approaches may be necessary depending on the location of the work zone, i.e. if work are performed on a central island and circulatory area is affected (case I.5 or I.6), a usage of safety cones may be enough to provide separation. An example (for work zones on a circulatory roads) has been presented in Fig. 1a and b. In Poland safety cones may be used only when works are quick-moving. For long-term works usage of traffic directing device U-21 is necessary to separate a part of the road for traffic flow from the closed part [13].

Similar solutions may be adopted for cases I.7 to I.14 when more than one element of a roundabout is affected by the work zone. An example, based on [12] has been presented in Fig. 2a and b. However it is essential to remember about providing minimal unobscured width of a traffic lane and to provide enough visibility for drivers.

It is also of great importance to inform drivers about narrowed traffic lanes and about location of work zone at a roundabout. Appropriate traffic signs should be used (i.e. A-14 in Polish conditions).

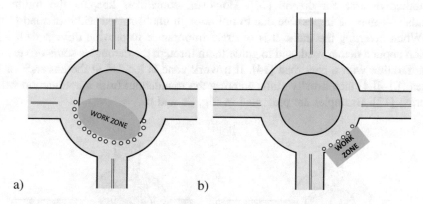

Fig. 1. Work zones on a circulatory road. (Source: own research based on [12])

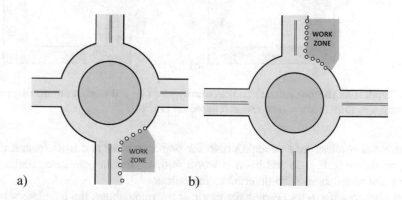

Fig. 2. Work zone affecting entrance and circulatory area (2a) and work zone affecting exit and circulatory area (2b). (Source: own research based on [12])

3.2   Cases Assigned to GROUP 2

Cases assigned to GROUP 2 are associated with full closure of a given element of roundabout: entrance, exit, circulatory area or combination of these elements at the same time. It is essential to try to minimize the time of works at a roundabout with full closure of an element.

When an element of roundabout is closed there are in general two ways to organize temporary traffic control [12, 14]:

- keep the traffic at the roundabout (i.e. using alternating traffic),
- divert some traffic from the roundabout.

Diverting some part of traffic may benefit in shorter time of works and increased safety for workers and drivers [14]. Nevertheless, there are some disadvantages of this solution, such as possible congestion on detour road or problem with travelling through an unknown area for drivers [14]. However, sometimes keeping the traffic at a roundabout may be impossible due to not enough unobstructed width of road [12].

When diverting the traffic it is of great importance to provide drivers with information about a detour road, and to guide them through the detour, as some drivers may be unfamiliar with a new road [14]. If a work zone is located at the entrance or exit (cases II.1–II.4) then usually traffic entering the roundabout from that direction will be diverted [12]. Examples are presented in Fig. 3a and b.

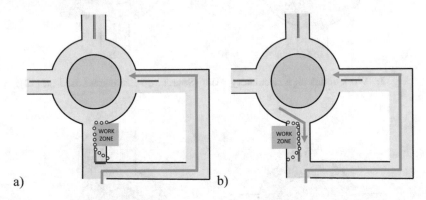

a)   b)

Fig. 3. Work zone affecting entrance and circulatory area (3a) and work zone affecting exit and circulatory area (3b). (Source: own research based on [12, 14])

Presented solution may be appropriate for both entrances and exits from a roundabout, as shown in Fig. 3a and b - it is worth noticing that in both cases traffic flows entering the roundabout were diverted to the detour.

It is also possible to try keeping the traffic at the roundabout. If a circulatory road is closed (cases II.5–II.6), then instead of diverting the traffic to a detour an alternating traffic may be used. However, it requires use of traffic control devices on each entrance in order to ensure that traffic from one entrance will enter the roundabout at a given moment and that the gap between traffic departing from the roundabout and entering will be long enough. Also, flaggers may be in charge of controlling the traffic [14].

An example of solution where circulatory road is closed and the traffic is kept at the roundabout has been presented in Fig. 4 with appropriate traffic signs. The distance from the roundabout to first traffic signs may depend on the type of the road (rural road, urban street) and the speed limit [14]. Although keeping the traffic on the roundabout means that drivers do not have to take a detour road, they may be unfamiliar with a new way of traffic flow when alternating traffic is in operation.

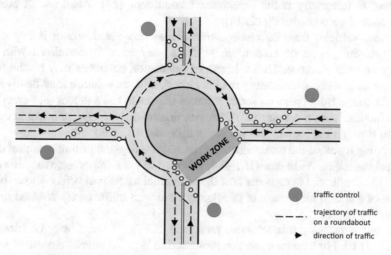

Fig. 4. Work zone on a circulatory road with traffic kept at the roundabout. (Source: own research based on [14])

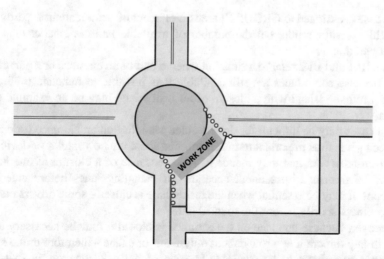

Fig. 5. Work zone on a circulatory road with traffic partly diverted. (Source: own research based on [14])

It is of great importance to inform drivers about changes in traffic organization Because each roundabout should be treated as unique location [12, 14], then it is necessary to adjust traffic control to specific features of each object. For example, sometimes an additional flagger may be needed on the central island to assist in traffic movement in temporary traffic organization conditions [14]. Also, use of protective vehicle should be considered [12, 14].

Because vehicles from only one entrance may enter at a time it is important to ensure that the period of time when vehicles may enter is correlated with traffic intensity on each entrance. Thus a length of queues on entrances may be shortened.

Sometimes, when a circulatory road is closed, a better solution may be diverting a part of the traffic from the roundabout. It forces drivers to use a detour (not every driver may be familiar with a detour road) however, drivers do not have to travel in opposite direction through the roundabout (as in solution shown in Fig. 4).

Diverting a part of the traffic also does not require flaggers or traffic control devices to control the traffic. As in cases II.1–II.4, it is also of great importance to guide drivers through the detour and to consider how the additional traffic may affect detour road. An example of solution for closure of circulatory road with traffic partly diverted has been presented in Fig. 5.

It is worth noticing that solutions presented in Figs. 4 and 5 may be adopted for cases II.7–II.10. For example, solution presented in Fig. 5 demands closure of entrance for traffic that is diverted - despite the fact that there is no work zone on the entrance.

3.3   Cases Assigned to GROUP III

When a case is assigned to GROUP III then an element of a roundabout is partly closed - it is still open for traffic, but the number of available lanes is smaller than during normal operation.

Cases III.1 and III.2 refer to a situation when a part of an entrance or a part of exit is closed. Because some lanes are still available it is possible to maintain traffic. Work zone on a refugee island with a closure of an inside lane may be an example of such situation.

Such cases may be difficult for drivers since visibility on an entrance (or exit) may be reduced - it is thus important to make sure that there are no vehicles or devices used for performing works that may reduce it. Also, because of a closure of one lane it is important to inform drivers about necessity of changing lanes before entering the roundabout. It may be essential when the inside lane is closed - some drivers may try to overtake others shortly before the roundabout.

Moreover, because one lane on the entrance is closed it may be necessary to allow driving in any direction at a roundabout only from one lane - therefore traffic signs on an entrance may have to be changed. In case of a work zone on an exit from a roundabout it is important to provide drivers with information about only one lane on the exit - some drivers may want to leave the roundabout directly from the circulatory road. Possible solutions are presented in Fig. 6a and b.

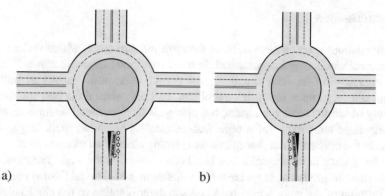

Fig. 6. Work zone on an entrance of a multi-lane roundabout (6a) and on an exit of a multi-lane roundabout (6b)

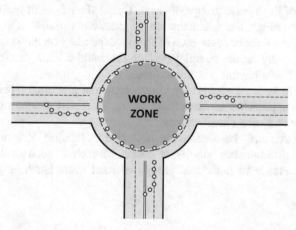

Fig. 7. Work zone on a circulatory road of a multi-lane roundabout. (Source: own research based on [14])

An element that is partly closed may also be the circulatory road - for example, the work zone on the central island may affect the circulatory road thus one lane must be closed for traffic. In case of a two-lane roundabout (as shown in Fig. 7) it is possible to maintain the traffic by transforming the roundabout to a single-lane roundabout. That may require closing inside lanes on entrances of a roundabout. Therefore it is of great importance to inform drivers about necessity of changing lanes before approaching the roundabout.

4   Conclusions

Each roundabout is an element of transportation infrastructure, where various types of planned road works may be performed. In most cases they require the use of temporary traffic organization. In such situations, the choice of optimal variant of traffic management is a very important decision problem. Not only should the best solution ensure the safety of workers and road users, but also guarantee minimal disruptions of traffic.

Variants of temporary traffic organization may be classified according to different criteria. In the article the number of available traffic lanes at an element of a roundabout during temporary traffic organization has been chosen as the main criterion. On this basis three main groups of cases have been distinguished. The additional consideration of the location of the work zone - based on the determination of the closed element (or elements) of the roundabout - allowed to develop a detailed typology of possible temporary traffic organization at the roundabout.

In addition, general conditions and selected examples of solutions have been presented for each of three basic groups of cases. Detailed analysis of individual cases for the developed typology and identification of possible variants of temporary traffic organization at the roundabout for each of them - depending on the existing constraints and conditions - may be an essential part of the multi-criteria decision model supporting optimal traffic planning in urban areas.

It is difficult to assess the frequency of occurrence of presented types of work zones due to different approaches to classifying them. Presented groups of cases may be useful to standardize such classifications.

Further research may be focused on objective evaluation of level of difficulties connected with distinguished groups of cases. Moreover, such evaluation may be conducted in reference to individual groups of road users (such as pedestrians, car drivers, etc.).

References

1. Brilon, W.: Roundabouts: a state of the art in Germany. http://techamerica.com/RAB14/RAB14papers/RAB14ppr045_Brilon.pdf
2. Wisconsin Department of Transportation: Roundabout Guide, Wisconsin (2008)
3. Ambros, J., Turek, R., Janoska, Z.: Safety evaluation of Czech roundabouts. Adv. Transp. Stud. Int. J. Sect. B **40**, 111–122 (2016)
4. Persaud, B.N., Retting, R.A., Garder, P., Lord, D.: Crash reductions following installation of roundabouts in the United States. Insurance Institute for Highway Safety, Arlington (2000)
5. Macioszek, E.: The road safety at turbo roundabouts in Poland. Arch. Transp. **33**(1), 55–67 (2015)
6. Transportation Research Board, U.S. Department of Transportation: Roundabouts: An Informational Guide, 2 edn. National Cooperative Highway Research Program. Transportation Research Board, Washington (2010)
7. Insurance Institute for Highway Safety. Highway Loss Data Institute: Roundabouts. Status Report. http://www.iihs.org/iihs/sr
8. Transportation Research Board: National Cooperative Highway Research Program. Synthesis 488. Roundabout Practices. Transportation Research Board, Washington (2016)

9. Żochowska, R.: Wielokryterialne Wspomaganie Podejmowania Decyzji w Zastosowaniu do Planowania Tymczasowej Organizacji Ruchu w Sieci Miejskiej. Oficyna Wydawnicza Politechniki Warszawskiej, Warszawa (2015)
10. Żochowska, R.: Improvement of traffic safety in road work zones. Logistyka - Nauka **4**, 3459–3467 (2014)
11. Żochowska, R.: Safety in work zone - practical aspects. Logistyka - Nauka **4**, 3469–3478 (2014)
12. Department for Transport: Safety at street works and road works a code of practice. Department for Transport, London (2013)
13. Zasady Oznakowania Miejsc Robót oraz Organizacji Ruchu na Terenie Oddziału GDDKiA w Opolu. https://www.gddkia.gov.pl
14. American Traffic Safety Services Association: Temporary traffic control for building and maintaining single and multi-lane roundabouts. American Traffic Safety Services Association, Fredericksburg (2012)

Method of Assessing Vehicle Motion Trajectory at One-Lane Roundabouts Using Visual Techniques

Ireneusz Celiński[✉] and Grzegorz Sierpiński

Faculty of Transport, Silesian University of Technology, Katowice, Poland
{ireneusz.celinski,grzegorz.sierpinski}@polsl.pl

Abstract. The main purpose of roundabouts is road traffic calming by triggering a vehicle motion trajectory change before entering the roundabout and consequential traffic deceleration when crossing the roundabout area. The article provides a discussion on results of research concerning vehicle motion trajectory at one-lane roundabouts. Specific cases have been studied to analyse the effect of one-lane roundabout geometry on vehicle motion trajectory. The relevant survey was conducted by video recording a road traffic scene at a roundabout. An image thus obtained made it possible to determine what is referred to as a dipped headlight track of vehicles by application of the Harris algorithm. The vehicle motion trajectories recorded in the survey were subject to statistical analysis in order to reveal any potential regularities that could be relevant to assessment of the current roundabout geometry. The measuring method proposed is more flexible compared to other vehicle motion trajectory recording techniques, as it enables vehicular traffic to be studied under most weather and illumination conditions, including at night, in non-illuminated road sections, and even in fog.

Keywords: Vehicle motion trajectories · Roundabout · Visual techniques
Road traffic safety

1 Introduction

The rationale behind creating roundabouts is road traffic calming, while at the same time, appropriate parameters of flow rate of road network cross-sections can be ensured. The goal of traffic calming is achieved by adequately designing the geometry of individual roundabout elements [1–8]. By appropriately developing the geometry of roundabout entries, it is possible to force road users to reduce driving speed to an extent that depends on the entry's geometrical parameters. With well-designed roundabout geometry, such driving speed reduction can reach up to several dozen per cent, thus contributing to a significant decrease in the number of accidents and reduction of their consequential severity [9–12].

Despite the numerous advantages of such a geometric arrangement of road intersections, there are specific dysfunctions that emerge in certain roundabouts. This may be the case of poorly designed entries, where one has failed in triggering considerable driving speed reduction. It can be observed wherever the available urban space used for extension of intersections is in shortage (roundabouts typically require more free space

© Springer Nature Switzerland AG 2019
E. Macioszek et al. (Eds.): Roundabouts as Safe and Modern Solutions in Transport
Networks and Systems, LNNS 52, pp. 24–39, 2019.
https://doi.org/10.1007/978-3-319-98618-0_3

than intersections of different traffic organisation). A situation one can frequently observe in roundabouts is that drivers of passenger cars and delivery vans make use of the roundabout apron to make the vehicle motion trajectory shorter. What also contributes to similar behaviours becoming increasingly typical is the growing number of SUVs in roads. Drivers of such cars use the apron to traverse the roundabout in a shorter route (while aprons are dedicated to traffic of a different type of vehicles).

Geometric parameters of roundabouts are designed to enable traffic of the largest vehicles, which causes that passenger cars and delivery vans, which account for more than 90% of all vehicles admitted to traffic in public roads, can resort to traffic trajectory varianting to a certain degree. Such varianting may affect road traffic safety. Real-life vehicle motion trajectories may be mapped using visual techniques by application of the dipped headlight position tracking method. The algorithm used in the research in question was based on the Harris method [13].

2 Field of Research

The survey addressed in this article was conducted at five roundabouts. Pictures of the junctions subject to examination have been provided in Fig. 1a–e. They regulate the traffic system in the centre of a small town with the population of 40 thousand.

Fig. 1. Pictures showing location of the roundabouts subject to the survey

Small distances between the objects of research exert mutually negative effect on their functioning. One can observe strong impact between them in the rush hour. Vehicle queues that form at entries to intersections no. 2, 3, 4 and 5 (see Fig. 1f) extend as long as to adjacent junctions on numerous occasions. According to the authors, where this is the case, one deals with accumulation of negative traffic phenomena due to how the roundabouts in question function.

The design geometry of the intersections subject to the research does not affect vehicles approaching from individual entries to such an extent that they are forced to change their motion trajectory, and consequently also driving speed, to a considerable degree. A roundabout entry of correct design and a central island that has been appropriately situated force driving speed reduction, since motion trajectory changing in the roundabout is necessary. The foregoing is not always the case in practice. In the course of the research, the authors observed the fact that, at all five intersections examined, there were entries where a decided majority of entering vehicles did not decelerate. In the rush hour, not only did it increase the risk of collision in road traffic, buy also caused quicker depletion of traffic capacity of the short street sections linking individual roundabouts located in a close distance from each other. As a consequence of this process, there was extensive rush hour congestion observed in the town centre. On account of this observation, motion trajectories of vehicles approaching individual roundabouts from selected entries were studied. For instance, the video recording of intersections no. 1 and 2 covered the motion trajectories marked with arrows in Fig. 2a and b. For the sake of the survey, the video camera recording vehicle motion trajectories was installed in front of the given entry under vehicle traffic surveillance.

In Fig. 3, the vehicle motion trajectory variants observed in real life have been plotted on one of entries to intersection no. 1 (variants A, B and C). Most drivers traverse the roundabout shown in Fig. 3 in a south-north axis by moving along the trajectory marked with the arrow labelled "B". On account of the existing roundabout geometry, this corridor does not trigger any significant driving speed reduction. One can virtually cut through this roundabout along the said corridor with the maximum permissible driving speed. And since this is practically the town's outbound road, the actual driving speeds tend to exceed the permissible ones, especially at night. SUVs often pass this roundabout in the trajectory marked "A", i.e. using the roundabout apron which is obviously intended for different purposes. Such a manoeuvre does not require any significant driving speed reduction as one traverses this corridors. In the travelling direction of vehicles moving along trajectories A, B and C, ca. 200 m past

a/

b/

Fig. 2. Pictures of selected roundabouts (no. 1 and 2) with trajectories marked

the roundabout, there is a pedestrian crossing which is mainly used by children heading towards a nearby primary school. Very few of the vehicles passing this roundabout use the trajectory marked "C" which should be considered optimum from the perspective of the roundabout's specific geometry. A travelling trajectory that would ensure driving speed reduction while crossing the roundabout is only used by vehicles approaching the intersection from other entries than the analysed one.

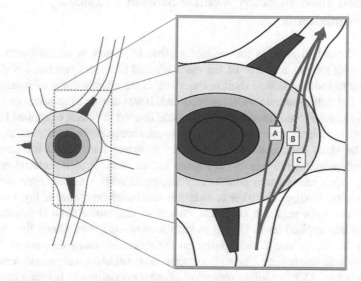

Fig. 3. Intersection no. 1 - vehicle motion trajectory variants observed in practice at one of entries (Source: own research based on [14])

Figure 4 shows the vehicle motion trajectory observed in intersection no. 2. Vehicles moving along the corridor marked with an arrow also abstain from reducing their speed to a significant degree. This is due to the intersection geometry. This case is

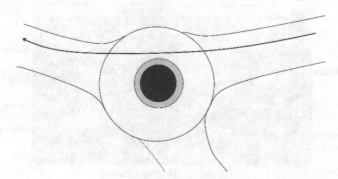

Fig. 4. Intersection no. 2 - vehicle motion trajectory observed in practice at one of entries (Source: own research based on [14])

particularly problematic on account of the close proximity of a rail bridge, under which one can frequently observe a traffic jam in the rush hour. The cases illustrated in Figs. 3 and 4 are also observed in intersections no. 3, 4 and 5. There is at least one entry at each of these intersections that - as vehicles enter the roundabout – causes no or minimum driving speed reduction.

3   Method Used to Study Vehicle Motion Trajectory at Roundabouts

The method used to analyse the vehicle motion trajectory at roundabouts has been discussed with reference to two of the cases studied (intersections no. 1 and 2).

In order to test the vehicle motion trajectory at the chosen entry of roundabout no. 1, the relevant traffic stream was video recorded. It was decided that some characteristic points of vehicles should be recorded, and these formed the track of dipped headlights. The dipped headlight is indeed one of the most characteristic elements in a vehicle outline. The video camera used in the survey recorded image with the resolution of 1280×960 pixels at the rate of 50 frames per seconds. It also featured an infrared filter. The video camera was additionally equipped with a visible light filter of the authors' original design. The filter in question was based on a double layer (not glued) of plastic with light-sensitive emulsion subject to standard chemical processing in a developing box applied on it. Owing to such a combination of filters, the elements of interest to the survey could be exposed on the recorded road traffic scene, and these were the vehicle headlights. The video camera also featured wide-angle lens with the visual field of ca. 150°, enabling reduction of what is commonly referred to as the fish eye effect. Each time, the video camera was installed in front of the entry being examined in such a manner that the angle between the camera focal length and the entry axis was as small as possible. This necessity resulted from the fact the main aspect of interest in the study was the vehicle deflection from the axis of the entry used to enter the roundabout.

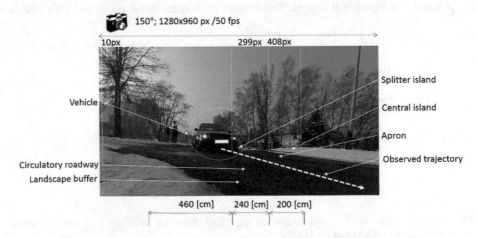

Fig. 5. Elements of the road traffic scene video recorded at intersection no. 1

Figure 5 shows the video camera view over the roundabout entry under examination (intersection no. 1). For the sake of increased legibility of the motion image presented in Fig. 5, the visible light filter has not been applied. The actual frame of the given road traffic scene, as perceived through the modified camera lens, has been provided in Fig. 6.

Fig. 6. Actual road traffic scene image recorded using both physical filters

Figure 6 shows that the physical filter installed on the camera lens causes that vehicle headlights, and the dipped headlights in particular, clearly stand out in road traffic scene. This enables implementation of an algorithm for the motion picture content analysis, also referred to as footage processing. In the course of further processing, the recorded traffic image was cropped to frame the road traffic scene to dimensions required for purposes of examination of the vehicle motion trajectory from the roundabout entry to the roundabout exit. Next, the image was transformed by means of a software median filter. The image analysis method applied used a headlight detection procedure based on the Harris algorithm [13]. By that means, the fact that dipped headlights are fixed points of reference in the vehicle outline made it possible to record motion trajectories of vehicles traversing the roundabout. This technique of examination enables analysis of vehicle motion trajectory under nearly any weather and illumination conditions. It proves efficient in non-illuminated out-of-town roads, in dense fog (when fog lights are video recorded) etc.

Figure 7 shows the main window of an original application used to measure and visualise the motion trajectory of vehicles traversing the roundabout. The application applies highly advanced footage processing techniques. On the left-hand side of the application window, the original image recorded by the video camera is displayed. The right-hand side of the application window (at the top) shows the image of the analysed traffic scene cropped out of the original frame with the resolution of 1280 × 960 pixels. The image framing operation is conducted on-the-fly, which enables minimisation of the number of image noises caused by objects found within the boundaries of the scene analysed, yet not involved in vehicular traffic. Once modified (converted and shrunk), the frame which represents the traffic scene is transformed by means of the median filter that reduces the number of erroneous measurement readouts. In the editing list displayed on the bottom right-hand side of the application window, one can view coordinates of the points identified using the Harris algorithm previously applied. These coordinates determine the positioning of dipped headlights of the vehicles within the frame.

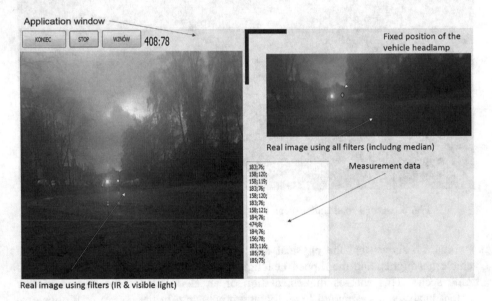

Fig. 7. Main window of the authors' original application used in the research

As the video footage was analysed, there were numerous problems connected with readout and interpretation of measuring points. They required an in-depth image analysis to retrieve correct measurement data. First and foremost, while vehicle motion trajectories were surveyed both during and after precipitation, not only the positioning of dipped headlights but also their reflections in the wet pavement were recorded (Fig. 8a). In some frames, two or more headlights of more than a single vehicle moving in different traffic streams were also recorded (Fig. 8b). These two cases (reflections in wet pavement and superposition of headlights from different traffic streams) may be eliminated by analysing the width of headlight spacing in various vehicles and vertical

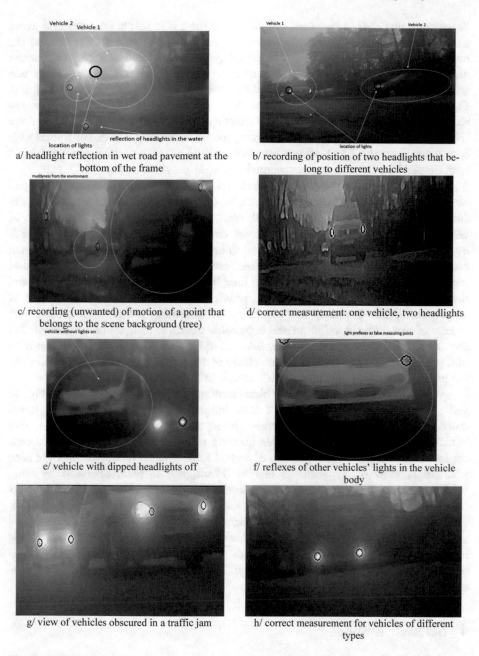

a/ headlight reflection in wet road pavement at the bottom of the frame

b/ recording of position of two headlights that belong to different vehicles

c/ recording (unwanted) of motion of a point that belongs to the scene background (tree)

d/ correct measurement: one vehicle, two headlights

e/ vehicle with dipped headlights off

f/ reflexes of other vehicles' lights in the vehicle body

g/ view of vehicles obscured in a traffic jam

h/ correct measurement for vehicles of different types

Fig. 8. Identification of measurement irregularities observed in the analysis of intersections no. 1 and 2

coordinates that characterise their positioning in the traffic scene. One can also avoid such issues by applying adequate settings of the camera's optical axis. Despite the complex filtering procedure applied in the analysed image, there were some incidental measuring points recorded, originating in the traffic scene environment, such as tree branches moving in the wind and causing changes to spot illumination (Fig. 8c). Figures 8b and c show that, in individual cases, positioning of only one headlight was recorded, which resulted from problems adapting coefficients of the image recognition (Harris) algorithm [13]. Such errors are reduced by comparing successive frames with each other. A correct frame with no interferences and with the position of the vehicle's both dipped headlights correctly recorded has been provided in Fig. 8d.

What is meant by correct measurement is recording of two points of dipped headlights for a single vehicle in a single frame (a motorcycle being a separate case). In such a frame, the elements marked with ovals should only be dipped headlights of vehicles, or alternatively other lights of the same vehicles (e.g. emergency services). Some measuring data irregularities related to background noise (vibration of different background objects, reflexes and reflections) are removed by analysing coordinates of measuring points, which proceeds in real time based on an analysis of the vertical and the horizontal coordinates of identified points recognised as potential positions of dipped headlights. Capturing and separation of measuring points that correspond to objects of two different traffic streams is possible thanks to analysis of chronologically sequenced data. Some frames showing the road traffic scene do not identify the position of dipped headlights, even though the vehicle is there. Such cases are compensated by the high rate of image recording which comes to at least 50 fps.

Another problem was posed by vehicles moving with dipped headlights off (Fig. 8e and f). Such vehicles are sources of false indications on account of light reflexes emerging on their bodies and produced by elements of the environment as well as other vehicles. Figure 8g illustrates vehicles queuing in a traffic jam. In the specific case depicted, identification of the position of dipped headlights is correct, however, in such a situation, view of characteristic objects (headlights) is often obscured by other vehicles, and the frame tends to be excessively illuminated, which may lead to errors on account of light reflexes on bodies of vehicles in close proximity. Figure 8h provides a frame showing correct measurement of the dipped headlight position in a vehicle other than a passenger car. In this case, a bus was video recorded in the roundabout's central carriageway. It was established that the vehicle type (also non-standard headlight contours) exerts no significant impact on the correctness of measurement data readout.

4   Selected Results

Figure 9 provides a graph of trajectories of several vehicles passing intersection no. 1.

This image encompasses vehicles recorded while moving along two different corridors, the trajectories of which should be deflected in order to trigger driving speed reduction near the roundabout's central island. On account of the specific position of the video camera's optical axis in the space, vehicle motion trajectories contain a reflection of the image in a three-dimensional space. The results imply that, although

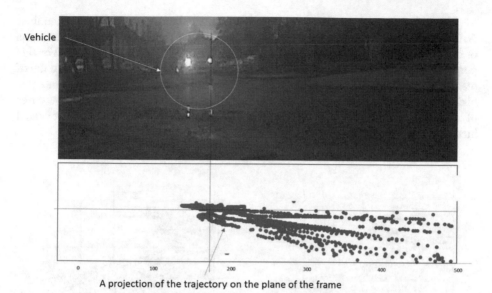

A projection of the trajectory on the plane of the frame

Fig. 9. Selected survey results for intersection no. 1

trajectories of some vehicles differ to a small extent from one another, none of the recorded vehicles were actually forced (due to the roundabout geometry parameters) to change their motion trajectories accordingly. Only those which approached the intersection from other destinations, going round the central island, were forced to decelerate.

Figure 10a, b and c illustrate vehicle motion trajectories recorded in a one-lane roundabout (intersection no. 2). They were trajectories recorded in different time intervals in a traffic corridor labelled "a/b/c" in Fig. 10f. Figures 10d and e show motion trajectories of vehicles moving along a corridor labelled "d/e" in Fig. 10f. In Fig. 10, the X axis represents the width of the recorded road traffic scene frame, while the Y axis corresponds to its height. The unit of measure of both axes is the number points in the image, known as pixels, the maximum number of which results from the resolution of the video camera used in the survey (1280 × 960). The vehicular traffic in both corridors demonstrated in Fig. 10 should involve a considerable change to the motion trajectory near the roundabout's central island (roughly at the middle of the height of axis Y for the plots in question). Contrary to this assumption, one cannot observe any significant trajectory change in Figs. 10d and e, and so the resultant driving speed reduction is also inconsiderable. This case is similar to the situation observed in intersection no. 1. Therefore, one may conclude that both intersections are in fact dysfunctional on a level of approximately 1/3. One entry out of three does not meet the requirements imposed upon such infrastructure elements as roundabouts. Due to fact that, in both intersection no. 1 and no. 2, the dysfunctional entries are used by predominant traffic streams, the factor of 1/3 must be considered as minimum. Trajectories of vehicle motion in roundabouts may be analysed by application of the method described above. Analyses such as those may be additionally performed in an automatic mode by means of machine teaching procedures.

The authors have formulated a hypothesis that whether or not vehicles reduce their driving speed at the roundabout's central island may be verified by statistical analysis of distribution of measuring points of individual vehicles' motion trajectories. For this purpose, the authors have calculated numerical characteristics of a set of data of measuring points shown in Fig. 10 (the traffic intensity observed in intersection no. 2 is higher than in other intersections), and for that purpose, they have compared two cases of a major motion trajectory change (Fig. 10a, b and c) with those that revealed inconsiderable trajectory change (Fig. 10d and e).

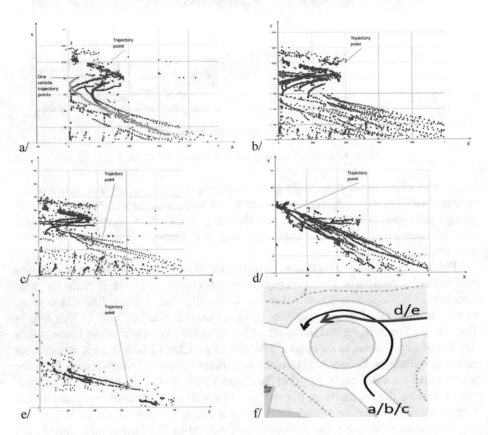

Fig. 10. Selected survey results for intersection no. 2

Based on the analysis of the values collated in Table 1, one may draw a conclusion that the fact of changing the profile of vehicle motion trajectory at the roundabout's central island is reflected in the numerical characteristics of the set of data describing the coordinates of points forming vehicle motion trajectories.

For a major vehicle motion trajectory change (and consequently also driving speed change), one can observe higher values of arithmetic mean, standard deviation and median for the Y coordinate which describes the points that form trajectories of

Table 1. Numerical characteristics of a set of data describing vehicle motion trajectories in intersection no. 2.

Numerical characteristics of data set*	Driving in an arc; major trajectory deflection - considerable driving speed reduction			Small trajectory deflection - inconsiderable driving speed reduction	
	a/	b/	c/	d/	e/
Mean X	109	125	134	125	114
Mean Y*	**69**	**66**	**55**	49	59
∂_y	83	92	99	95	99
∂_{y_*}	**28**	**31**	**36**	22	19
Med_X	106	116	125	102	88
Med_Y*	**74**	**78**	**60**	49	60

Values in the table have been rounded to an integer.

individual vehicles. With reference to the roundabout cases studied, the X coordinate does not provide a good description of these changes. Especially the value of the mean and the median for the Y coordinates of measuring points is statistically significant for distinguishing between cases of vehicles passing the roundabout with and without driving speed change. In Table 1, for columns a, b and c, values of the mean and the median provided are also considerably understated on account of the measurement of positions of some vehicles moving along corridors d and e (the video camera captured headlights of a part of these vehicles).

Having statistically analysed the measuring points which represent the dipped headlights of vehicles, in accordance with the method proposed, one can assess the given roundabout's geometry. Similar statistical relationships were essentially observed at each of the five intersections surveyed. Observation of the vertical coordinate of measuring points (Y) made it possible to detect anomalies in motion trajectories for these intersections. In each of them, at least one of the examined entries did not cause any major trajectory change, and consequently also no vehicle driving speed change as well. The method in question requires further in-depth studies and analyses conducted on a larger sample of intersections and in a higher number of measurement periods. This applies to night time and weekends in particular, since roundabouts of inappropriate geometry tend to favour drivers inclined to exceed the permissible driving speed.

5   Analysis of Irregularities and Measurement Error

On account of the traffic stream recording technique applied and the use of automatic footage processing algorithms, the method proposed by the authors is heavily exposed to measurement errors. The main interfering factors are moving objects within the frame that do not belong to vehicles (trees, people, light reflexes etc.). Further errors are related to measuring the position of only one headlight or the view of headlights being obscured by vehicles preceding others in the traffic stream. To a certain extent, these errors can be eliminated by correct framing and by using appropriate physical and software filters. For that reason, assuming that only the position of dipped headlights is recorded in each frame, every odd occurrence of the number of objects identified in the given image (measurement points) is to be interpreted as erroneous. Graphical illustration of the interpretation error for successive frames has been provided in Fig. 11.

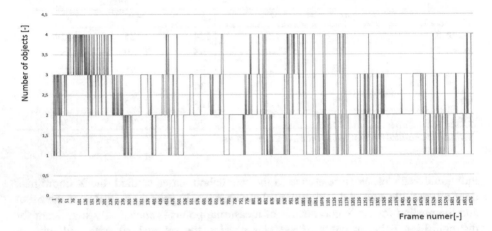

Fig. 11. Number of objects (vehicle headlights) identified in specific time in intersection no. 2 reflecting the potential algorithm interpretation errors

In this figure, every odd number corresponds to a case of the view of one headlight of the vehicles video recorded in the frame being obscured or to missing identification of such a headlight. The latter case is relatively rare, as the predominant problem is the view of headlights being obscured by other vehicles. The study may also include cases of video recording of vehicles moving with only one headlight operational. The measurement illustrated in Fig. 12 revealed the following percentage breakdown: 22.97% of frames with one measuring point, 44.70% of frames with two measuring points (one vehicle), 20.66% of frames with three measuring points and 11.66% of frames with four measuring points (two vehicles). The measuring video camera was installed in such a manner that the number of vehicles recorded in a single frame would be as low as possible.

Cases of video recording of one or three measuring points constituted the initial interpretation error for successive frames, resulting from the measuring method specificity, and in this case it accounted for ca. 43% of the recorded frames (this not being a measurement error typical of the method itself). For the objects examined, the rates of such errors ranged from 31.40% to 47.02%. Frame interpretation errors may be rectified by modifying the video camera's optical axis, but they can mainly be minimised in the video stream analysis, as successive frames are being compared (a motion sequence is recorded chronologically, with accuracy down to 50 fps), which makes it possible to determine the position of lost indirect measuring points (the case of recording 1 or 3 measuring pints). What proves helpful in the measurement data

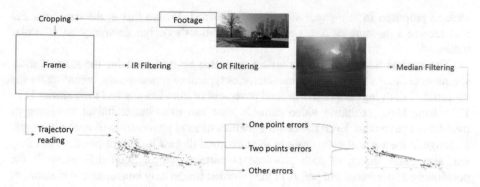

Fig. 12. Image filtering and measurement data adjustment

adjustment procedure is the symmetric arrangement of dipped headlights against the vehicle's vertical axis. Consequently, the actual measurement error was observed to range from 8.05% to 14.92%. The pattern of image filtering and measurement data adjustment has been illustrated in Fig. 12.

6 Conclusions and Further Research

The method applied in the research enables analysis of motion trajectories of vehicles passing roundabouts. On account of the main goal of these intersections, namely road traffic calming and temporary reduction of driving speed, such a method may provide grounds for the roundabout geometry assessment. The preliminary studies conducted by the authors have enabled them to establish as a fact that all the one-lane roundabouts they have examined do not fulfil their fundamental purpose to a certain extent. At least one entry of each of the intersections subject to the survey is improperly designed, enabling the roundabout to be passed with either no or merely a minimum driving speed change. This pertains to the corridors characterised by the highest traffic flow. Not only does it pose potential threats in road traffic, but also with regard to the traffic system observed in the area covered by the analysis, where the intersections in question are located, it gives rise to spatially extensive traffic jams.

Using a wide-angle video camera in the survey, it is possible to establish the trajectory of motion in roundabouts with more than one traffic lane around the central island. In this case, however, it requires more cameras to be used and appropriate framing perspective to be set. This makes it possible to investigate other roundabout characteristics as well, including weaving in traffic. The research in question has not taken one-track vehicles into consideration, which should be rectified in the course of further studies on account of the research results addressed by other authors in the literature of the subject [15, 16], as dependence between roundabout solution and emission volume or traffic of one-track vehicles exerts considerable influence on accident rates typical of these intersections.

Since one of the research deliverables is a certain representation of a route-time graph, it is possible to determine the speed of vehicles circulating around the round-about's central island. This constitutes an important orientation towards which the method proposed in this paper should be developed. Where this is the case, one can also receive a measure of direct effect of roundabout's design geometry on vehicular traffic.

The method addressed in this article may also be perfected for the sake of measurement of other vehicle traffic characteristics typical of roundabouts, including for the potential of studying parameters of road pavement in roundabouts and their entries [17–19]. Using high resolution video cameras, one can examine technical condition of roundabout pavement. Even minor irregularities of road pavement will not only trigger a change in the vertical coordinate of the measured dipped headlight position, but they will also cause fading of such positions in certain frames. Such differences in the positioning of measurement points in the recorded image may constitute the measure of the road pavement condition.

What the authors noticed while surveying intersection no. 1 was some interesting characteristics of the process of vehicles merging with traffic from a stop in parking areas located in a direct vicinity of a roundabout entry. A roundabout of adequately designed geometry situated next to a large parking area exerts a major impact on traffic calming with regard to vehicles attempting to merge with traffic [20]. A route previously planned in the road network, based on chains of random intermediate points resulting from the scatter of parking spots may also lead to a change in motion trajectory. It may be important to include that kind of analysis to traffic flow simulation on roundabouts [21].

References

1. Federal Highway Administration: Roundabouts: An Informational Guide. Publication No. FHWA-RD-00-067. Kittelson & Associates, Portland, Oregon (2000)
2. Mills, A., Duthie, J., Machemehl, R., Waller, T.: Texas Roundabout Guidelines. CTR, Austin (2011)
3. Department of Transport (United Kingdom): Geometric Design of Roundabouts. http://www.standardsforhighways.co.uk/ha/standards/dmrb/vol6/section2/td1607.pdf
4. Brilon, W., Bondzio, L.: Untersuchung von Mini-Kreisverkehrsplaetzen. Ruhr-University, Bochum (1999)
5. Macioszek, E.: The comparison of models for critical headways estimation at roundabouts. In: Macioszek, E., Sierpiński, G. (eds.) Contemporary Challenges of Transport Systems and Traffic Engineering. LNNS, vol. 2, pp. 205–219. Springer, Switzerland (2017)
6. Szczuraszek, T., Macioszek, E.: Proportion of vehicles moving freely depending on traffic volume and proportion of trucks and buses. Baltic J. Road Bridge Eng. 8(2), 133–141 (2013)
7. Pilko, H., Brčić, D., Šubić, N.: Speed as an element for designing roundabouts. In: Proceedings of 2nd International Conference on Road and Rail Infrastructure - CETRA 2012, pp. 981–988. CETRA Press, Dubrovnik (2012)
8. Šubić, N., Legac, I., Pilko, H.: Analysis of capacity of roundabouts in the city of Zagreb according to HCM-C-2006 and Ning Wu methods. Tech. Gaz. 19(2), 451–457 (2012)
9. Legac, I., Ključarić, M., Blaić, D., Pilko, H.: Traffic safety of roundabouts in the City of Zagreb. Proc. Med. Tech. Legal Aspects Traffic Saf. 1, 49–55 (2009)

10. Macioszek, E.: Analiza Prędkości Przejazdu Pojazdów Przez Skrzyżowania z Ruchem Okrężnym. Systemy, Podsystemy i Środki w Transporcie Drogowym, Morskim i Śródlądowym. Prace Naukowe Politechniki Warszawskiej Seria Transport, vol. 82, pp. 69–84 (2012)
11. Brude, U., Larsson, J.: What roundabout design provides the highest possible safety? Nordic Road Transp. Res. **2**, 17–22 (2000)
12. Harper, N.J., Dunn, R.C.M.: Accident prediction models at roundabouts. In: Proceedings of the 75th ITE Annual Meeting, Institute of Transportation Engineers, Washington, pp. 1–15 (2005)
13. Harris, C., Stephens, M.: A Combined Corner And Edge Detector. https://www.cis.rit.edu/~cnspci/references/dip/feature_extraction/harris1988.pdf
14. Overpass Service. https://overpass-turbo.eu/
15. Coelho, M., Farias, T.L., Rouphail, N.: Effect of roundabout operations on pollutant emissions. Transp. Res. Part D Transp. Environ. **11**(5), 333–343 (2006)
16. Macioszek, E., Sierpiński, G., Czapkowski, L.: Methods of modeling the bicycle traffic flows on the roundabouts. In: Mikulski, J. (ed.) Transport Systems Telematics. CCIS, vol. 104, pp. 115–124. Springer, Heidelberg (2010)
17. Staniek, M.: Stereo vision method application to road inspection. Baltic J. Road Bridge Eng. **12**(1), 38–47 (2017)
18. Staniek, M.: Road pavement condition as a determinant of travelling comfort. In: Sierpiński, G. (ed.) Intelligent Transport Systems And Travel Behaviour. AISC, vol. 505, pp. 99–107. Springer, Cham (2017)
19. Staniek, M.: Moulding of travelling behaviour patterns entailing the condition of road infrastructure. In: Macioszek, E., Sierpiński, G. (eds.) Contemporary Challenges of Transport Systems And Traffic Engineering. LNNS, vol. 2, pp. 181–191. Springer, Cham (2017)
20. Pypno, C., Sierpiński, G.: Automated large capacity multi-story garage - concept and modeling of client service processes. Autom. Constr. **81**(1), 422–433 (2017)
21. Małecki, K., Wątróbski, J.: Cellular automaton to study the impact of changes in traffic rules in a roundabout: a preliminary approach. Appl. Sci. **7**(7), 742 (2017)

Road Pavement Condition Assessment at Selected Roundabouts in the Town of Tychy

Marcin Staniek[✉]

Faculty of Transport, Silesian University of Technology, Katowice, Poland
marcin.staniek@polsl.pl

Abstract. The purpose of the study is to identify technical condition of road pavement at selected roundabouts in the town of Tychy by means of the RCT tool. The solution proposed for this purpose is a component of an integrated IT system referred to as the S-mileSys platform, intended for freight transport support in urban areas using information and communication technologies (ICT). The measuring method applied in RCT is essentially based on analysis of the relationship between linear accelerations recorded while vehicles move in the given road network elements. The research elaborated in this paper covered twelve roundabouts whose pavement condition was identified in a breakdown into roundabout carriageways, entries and exits. The results obtained in measurements of traffic dynamics make it possible to estimate indicators of the road infrastructure condition assessment, and consequently also to indicate roundabouts in need of more detailed examinations or planned repairs, either local or more comprehensive.

Keywords: Road pavement diagnostics · Identification of road defects
Pavement structure · Analysis of linear accelerations

1 Introduction

As described in the literature of the subject, a roundabout is commonly understood as an at-grade intersection of roads where traffic has been organised around the central island and, depending on the given country's legal regulations, it proceeds in the counterclockwise direction for the right-hand traffic and in the clockwise direction for the left-hand traffic. The roundabout's main function is to reduce the speed of traffic and ensure adequate visibility. Compared to road junctions of other types, i.e. intersections where vehicle traffic trajectories collide, roundabouts are characterised by generally superior flow rates and more undisturbed traffic flow. Owing to their design, roundabouts minimise weaving of traffic streams, which significantly increases the intersection safety levels. One of roundabout variations is a turbo roundabout where the vehicle motion trajectory is collisionless starting from the vehicle's entry until it exits the roundabout. The motion trajectory path itself depends on the exit choice, i.e. on the lane the driver uses to enter the given intersection [1].

Where it is justified, intersections feature linking sections that allow for the roundabout's circulatory carriageway to be by-passed. One can also come across numerous alternative solutions which, for instance, enable buses and lorries to cut

© Springer Nature Switzerland AG 2019
E. Macioszek et al. (Eds.): Roundabouts as Safe and Modern Solutions in Transport
Networks and Systems, LNNS 52, pp. 40–49, 2019.
https://doi.org/10.1007/978-3-319-98618-0_4

through the central island, since on account of the size as well as the turning radius of these vehicles they could not cross an intersection otherwise, or they would but to the expense of considerable traffic disturbance. The TLS sector is particularly important [2–5]. Roundabouts of the largest diameters include yet another kind of design solutions, namely individual features being opened periodically and enabling transport of oversize cargo. Engineering solutions used in small and mini-roundabouts allow vehicles of higher categories to traverse directly through the central island. In many cases, the central islands are simply painted on the circulatory carriageway, thus enabling unobstructed passage of vehicles.

What is required to ensure high level of traffic safety in circular intersections is, first and foremost, driving speed reduction before the roundabout entry, and secondly, reduction of driving speed of vehicles traversing the circulatory carriageway. With regard to the latter, the driving speed reduction is possible by bending traffic tracks, introducing entries and exits of small radius curvature as well as small widths, thus triggering what is referred to as traffic throttling. Safety priorities can be managed by controlling traffic obstructions at roundabouts, for instance, by elevating the central island over the pavement surface, Ensuring good visibility both at the entry and within the roundabout itself as well as by properly managing pedestrian traffic, e.g. by forming splitter islands at roundabout entries.

Besides the safety factors, what also matters is the roundabout's flow capacity and efficiency. Every structure should ensure that all types of vehicles admitted to road traffic can actually pass the roundabout [6, 7]. For this purpose, one must first define a reference vehicle for which dedicated geometrical and organisational traffic solutions should be introduced so that it causes no traffic disturbance and conforms with respective design guidelines applicable to such road features. What also matters is that the roundabout flow capacity should be higher than intensity of traffic streaming towards the intersection at the reference hour, assuming that temporary traffic volume variations are permissible. It is relevant for the capacity estimation purposes that one calculates both the actual and the peak capacity. The literature of the subject describes the state of efficiency of roundabouts as a condition where it is possible to ensure flow capacity higher that traffic intensity and where passage tracks are adapted to driving speeds ranging at 20–30 km/h. There are also simulation tools to assess the efficiency of the solution [8, 9].

Next to the functional advantages addressed in papers [10, 11], roundabouts also involve certain restrictions. One should not introduce such solutions in arterial roads with a privileged traffic direction, e.g. with traffic light coordination, or where a tram line is routed without traffic lights [12]. An issue typical of large-diameter roundabouts involves long pedestrian and bicycle crossings being detrimental to traffic safety and decisive of acceptance for such solutions. The choice of a roundabout type always depends on traffic size and vehicle composition, directional distribution of traffic, surrounding structures as well as space limitations, pedestrian and bicycle traffic intensity, and it is established on a case-by-case basis [13, 14].

2 Road Pavement Diagnostics Based on Vehicle Motion Dynamics

Not only are specific geometric requirements imposed upon circular intersections defined in the literature of the subject, but also guidelines for design of the roundabout pavement structure. There is a catalogue of typical structures of flexible and composite pavement. Under the intersection engineering procedure, the traffic category should be chosen as if one designed the pavement structure of the entry most intensely encumbered with traffic. Regardless of the pavement structure solution assumed in the design and the forces affecting the pavement due to the traffic of vehicles, including heavy vehicles, it is required that road pavement be properly maintained, its condition monitored and all the necessary repair and renovation works conducted.

There are many methods for assessment of road pavement condition, from manual assessment performed by qualified road administration inspectors to more complex solutions based on highly advanced image processing technologies and laser scanning systems. The technological progress observed in terms of development of electronic equipment and innovative solutions supports road condition diagnostics, without which adequate road infrastructure management would be virtually impossible. There are popular and widely used tools for assessment of road pavement load capacity and identification of the layout of road courses, including voids in the pavement structure. Vehicles equipped with a wide range of measuring devices, such as laser surface analysers or image processing systems, perform high-precision measurements of pavement condition and identify parameters which describe road pavement degradation. One can use popular systems for identification and assessment of anti-slip properties, but there are also new solutions that enable high-scale mapping of pavement structure.

Among such road infrastructure condition assessment solutions, there are tools based on analysis of linear accelerations which describe vehicle motion dynamics of traffic in the road network [15, 16]. The basic principle is that a data collector records accelerations caused by vehicle motion and affected by such factors as the road infrastructure condition. Such a system's architecture has been described in [17], while its main components are the data collector, a car mounted terminal and a processing system. The results obtained by this method indicate the relative error of the system proposed at <10% compared to laser scanning technologies. Such high precision of the method proves accuracy, efficiency and reliability of this solution. What one can observe in the literature of the subject is a trend to filter out noises resulting from the vehicle mass damping [18, 19]. An interesting solution has been described in paper [20], namely that the filtering process is based on a vehicle motion description derived from drive tests performed over a portable hump with a known size. The transfer function thus obtained, established in the process of the half-car model simulation, reflects the vehicle's tilts. Paper [21] provides confirmation of the linear dependence of the accelerations recorded (taking average speed into account) on the road pavement unevenness.

The research in question, comprising assessment of pavement condition in selected roundabouts in the town of Tychy, was performed by means of an original and implemented tool for recording and analysis of linear acceleration signals known as

Road Condition Tool (RCT), being one of project components deployed under development of the S-mileSys platform. One of the platform's main goals is transport route planning over the first/last mile by taking the technical condition of road infrastructure into consideration. The idea behind the entire S-mileSys system is pursued under the international project entitled "Smart platform to integrate different freight transport means, manage and foster first and last mile in supply chains (S-mile)" under the "Sustainable Logistics and Supply Chains" call within the framework of the ERANET Transport III programme [22]. There are six institutions involved in the project, including businesses and higher technical schools from Spain, Turkey and Poland. The RCT tool arose from collaboration of scientists representing the Silesian University of Technology as a result of a review of available solutions for analysis and assessment of road pavements, their complexity and simplicity of the relevant measurement procedure.

2.1 RCT's Measuring System

The tool developed for data acquisition for the purposes of the road infrastructure condition assessment is a mobile device running on the Android operating system and with the RCT mobile application developed under the S-mile project on board. The device performs high-frequency recording of linear accelerations obtained from a MEMS module and links them with individual locations identified with reference to positions acquired from a GPS receiver. All the data are stored in the device's memory, and once the given route is completed, sent to the S-mileSys platform's data server via the GPS module. The RCT mobile application's architecture, user interface and features have been described in [23]. An example of the measurement procedure along with the application window has been shown in Fig. 1.

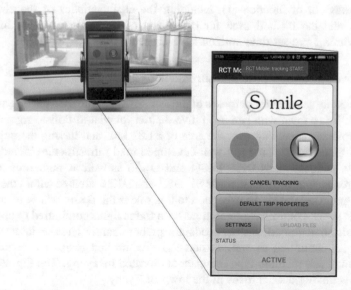

Fig. 1. Measurement procedure for road pavement condition assessment

After the signals have been recorded and the vehicle motion dynamics described, and once the data have been uploaded to the RCT server, the relevant data are verified and processed in series. What the system checks is the data series continuity and correctness of the set of values defining both accelerations and locations. The discontinuities identified in the data series in the function of time as well as the recognised deviations initiate a filtering procedure. What follows this stage is conversion of a route recorded as a series of GPS coordinates into a set of consecutive OSM map sections [24]. Next, it is possible to determine the road pavement condition assessment index for the given map sections, which is done according to the following relationship:

$$\delta\left(d_{id\_part}\right) = \left[1 + e^{-\beta(d-T)}\right]^{-1} \tag{1}$$

where:

d - sum of the absolute differences of the set of linear accelerations given by formula (2), where the set size depends on measuring frequency and test vehicle's driving speed,

β - parameter defining the function curve slope, where a high value implies stepwise behaviour of the function,

T - cut-off threshold for the vehicle's free vibrations

$$d_{id\_part} = \sum_{j} \left| a_j^y - a_{j-1}^y \right| \tag{2}$$

where:

a^y - recorded vertically oriented linear accelerations assigned to the given road section.

The behaviour of function (1), including the characteristics of the choice of its parameters and the method used for estimation of assessment indices for multiple transfers within a pre-set time horizon have all been discussed in [23].

2.2 Characterisation of the Measurement Area

The municipality selected for purposes of the assessment of the road pavement condition was Tychy. It is a town with administrative district (powiat in Polish) rights located in Silesian Province, stretching over the area of 81.81 km^2 and having the population of nearly 130 thousand. It features a well-developed road infrastructure including the S1 expressway, national roads DK1, DK44 and DK86 as well as numerous district and municipal roads. Intersections in roads S1, DK1 and DK86 situated within the borders of Tychy are grade separated. An exception to this rule is the DK44 whose main junctions are at-grade signal-controlled intersections with traffic lights configured to operate on an acyclic basis. Junctions of district roads are predominantly circular intersections and regular signal-controlled as well as non-signal-controlled ones. There are currently nineteen small and medium-type roundabouts located in Tychy. The Fig. 2 highlights roundabouts linking district roads in the town of Tychy.

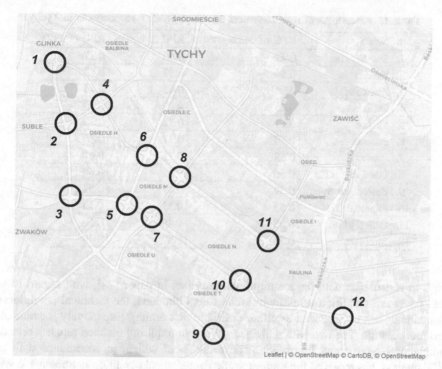

Fig. 2. Roundabouts subject to road pavement assessment

3   Assessment of Pavement Condition at Roundabouts

The pavement condition assessment process addressed in this paper was conducted at selected roundabouts over five consecutive workdays starting from 4 December 2017 in time of non-peak traffic volume. The data describing road pavement condition were recorded during multiple passes made with a passenger car featuring the RCT mobile data recording tool on board. All measurements were taken by the same driver who maintained constant driving speed while traversing the roundabouts. Following the measurements, the recorded data of the motion dynamics description were uploaded to RCT server in order to estimate the relevant road pavement condition indices.

By extracting road pavement condition indices for individual roundabout pavement parts as well as roundabout entries and exits from the database, it was possible to analyse the verified results that described the road infrastructure. Table 1 provides mean values of the pavement condition assessment indices in a breakdown into roundabout's structural elements. The mean values have been calculated for all pavement condition assessment indices with respect to individual roundabout elements. The Table additionally contains information concerning roundabout elements of poor technical condition, as per the RCT tool assessment.

The above analysis implies that with regard to the 12 selected roundabouts located in the town of Tychy, the average values of the road pavement condition assessment index calculated for roundabout carriageways, entries and exits range between 0.29 and

Table 1. Assessment of road pavement condition RPC of roundabouts in Tychy.

No	Name	RPC of carriageways	RPC of entries	RPC of exits
1.	Św. Jadwigi	0.43	0.60	0.37
2.	Harcerskie	0.56	0.32	0.52
3.	Żwakowskie	0.36	0.42	0.46
4.	Sublańskie	0.25	0.29	0.33
5.	Skałka	0.61	0.71	0.73
6.	Polonia	0.50	0.37	0.35
7.	Lwowskie	0.58	0.71	0.61
8.	Cassino	0.56	0.41	0.60
9.	Zesłańców Sybiru	0.39	0.47	0.53
10.	Paprocańskie	0.55	0.35	0.43
11.	Olimpijskie	0.53	0.49	0.56
12.	Strefowe	0.56	0.52	0.49

0.68. In accordance with the assumptions described in paper [24], with regard to the road pavement condition assessment indices thus obtained, the technical condition of roundabouts has been graded as either desirable or warning, respectively for nine and two roundabouts. The two with damaged pavement requiring planned repair works are roundabouts Skałka and Lwowskie. The roundabout which, in accordance with the RCT method, has received the highest grade for technical condition is Sublańskie with the roundabout pavement condition assessment index of 0.25, while these of individual entry branches are: 0.28 for *North* entry, 0.23 for *East* entry, 0.31 for *South* entry, 0.33 for *West* entry, and analogically of exits: 0.35 for *North* exit, 0.29 for *East* exit, 0.27 for *South* exit, 0.41 for *West* exit.

Having analysed these results, one can clearly establish that values of the round-about pavement condition assessment indices for the circular carriageway, the entry and the exit are independent of each other. One can also tell the difference between values of indices for individual roundabout entries and exits. This is particularly evident at roundabouts Harcerskie and Paprocańskie, whose condition assessment indices differ by 0.35 at entries *East* and *South* and by 0.29 at exits *North* and *South*. Among the potential reasons for increased degradation of road pavement at selected entries and exits, the most probable one is the number of vehicles moving in specific directions and the traffic composition, including intensity of heavy vehicle traffic through the given roundabout.

4   Conclusions

The assessment of road pavement condition of selected roundabouts in Tychy was conducted by means of the RCT tool, previously designed and implemented. It enables efficient recording and analysis of linear acceleration signals in order to identify and document the current condition of road infrastructure. From the perspective of local authorities and road administration bodies operating on the municipal level, the RCT tool provides support for road pavement management and maintenance. It also allows for monitoring of the road network's operating parameters by way of measurements causing no disturbance to the infrastructure and traffic participants. The pavement condition assessment results analysed in this paper with regard to individual roundabout elements have confirmed the relevant visual assessment.

Further research will comprise analysis of dependence between deterioration of technical condition of roundabout elements and the number of vehicles using the roundabout circular carriageway to travel in individual directions by taking the vehicle composition of traffic, including heavy vehicles, into consideration. Vehicle motion trajectories at the roundabout itself and in linking sections will be identified in order to analyse the correlation between pavement degradation at the roundabout as well as at entries and exits.

Acknowledgements. The present research has been financed from the means of the National Centre for Research and Development as a part of the international project within the scope of ERA-NET Transport III Sustainable Logistics and Supply Chains Programme "Smart platform to integrate different freight transport means, manage and foster first and last mile in supply chains (S-mile)".

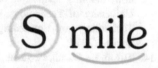

References

1. Macioszek, E.: The application of HCM 2010 in the determination of capacity of traffic lanes at turbo roundabout entries. Transp. Prob. **11**(3), 77–89 (2016)
2. Turoń, K., Golba, D., Czech, P.: The analysis of progress CSR good practices areas in logistic companies based on reports "Responsible Business in Poland. Good Practices" in 2010–2014. Sci. J. Sil. Univ. Technol. Ser. Transp. **89**, 163–171 (2015)
3. Golba, D., Turoń, K., Czech, P.: Diversity as an opportunity and challenge of modern organizations in TSL area. Sci. J. Sil. Univ. Technol. Ser. Transp. **90**, 63–69 (2016)
4. Galińska, B.: Multiple criteria evaluation of global transportation systems - analysis of case study. In: Sierpiński, G. (ed.) Advanced Solutions of Transport Systems for Growing Mobility. AISC, vol. 631, pp. 155–171. Springer, Cham (2017)
5. Żak, J., Galińska, B.: Multiple criteria evaluation of suppliers in different industries-comparative analysis of three case studies. In: Żak, J. (ed.) Advanced Concepts, Methodologies and Technologies for Transportation and Logistics. AISC, vol. 572, pp. 121–155. Springer, Cham (2017)
6. Macioszek, E., Sierpiński, G., Czapkowski, L.: Problems and issues with running the cycle traffic through the roundabouts. In: Mikulski, J. (ed.) Transport Systems Telematics. CCIS, vol. 104, pp. 107–114. Springer, Berlin (2010)

7. Macioszek, E., Sierpiński, G., Czapkowski, L.: Methods of modeling the bicycle traffic flows on the roundabouts. In: Mikulski, J. (ed.) Transport Systems Telematics. CCIS, vol. 104, pp. 1115–1124. Springer, Berlin (2010)
8. Małecki, K.: Graph cellular automata with relation-based neighborhoods of cells for complex systems modelling: a case of traffic simulation. Symmetry 9(12), 322 (2017)
9. Małecki, K., Wątróbski, J.: Cellular automaton to study the impact of changes in traffic rules in a roundabout: a preliminary approach. Appl. Sci. 7(7), 742 (2017)
10. Macioszek, E., Czerniakowski, M.: Road traffic safety-related changes introduced on T. Kościuszki and Królowej Jadwigi streets in Dąbrowa Górnicza between 2006 and 2015. Sci. J. Sil. Univ. Technol. Ser. Transp. 96, 95–104 (2017)
11. Macioszek, E.: Analysis of significance of differences between psychotechnical parameters for drivers at the entries to one-lane and turbo roundabouts in Poland. In: Sierpiński, G. (ed.) Intelligent Transport Systems and Travel Behaviour. AISC, vol. 505, pp. 149–161. Springer, Switzerland (2017)
12. Małecki, K., Pietruszka, P., Iwan, S.: Comparative analysis of selected algorithms in the process of optimization of traffic lights. In: Nguyen, N., Tojo, S., Nguyen, L., Trawiński, B. (eds.) Intelligent Information and Database Systems. LNCS, vol. 10192, pp. 497–506. Springer, Cham (2017)
13. Okraszewska, R., Nosal, K., Sierpiński, G.: The role of the Polish universities in shaping a new mobility culture - assumptions, conditions, experience. Case study of Gdansk University of Technology, Cracow University of Technology and Silesian University of Technology. In: Proceedings of ICERI 2014 Conference, pp. 2971–2979. ICERI Press, Seville (2014)
14. Turoń, K., Czech, P., Juzek, M.: The concept of walkable city as an alternative form of urban mobility. Sci. J. Sil. Univ. Technol. Ser. Transp. 95, 223–230 (2017)
15. Młyńczak, J., Burdzik, R., Celiński, I.: Research on vibrations in the train driver's cabin during maneuvers operations. In: Awrejcewicz, J., Kaźmierczak, M., Olejnik, P., Mrozowski, J. (eds.) 13th Conference on Dynamical Systems Theory and Applications. Abstracts DSTA 2015, p. 218. Łódź university of Technology, Łódź (2015)
16. Burdzik, R., Celiński, I., Czech, P.: Optimization of transportation of unexploded ordnance, explosives and hazardous substances-vibration issues. Vibroeng. Procedia 10, 382–386 (2016)
17. Du, Y., Liu, C., Wu, D., Li, S.: Application of vehicle mounted accelerometers to measure pavement roughness. Int. J. Distrib. Sens. Netw. 12(6), 1–8 (2016)
18. Ghose, A., Biswas, P., Bhaumik, C., Sharma, M., Pal, A., Jha, A.: Road condition monitoring and alert application: using in-vehicle smartphone as internet-connected sensor. In: Pervasive Computing and Communications Workshops - IEEE International Conference, pp. 489–491. IEEE Press, Lugano (2012)
19. Agostinacchio, M., Ciampa, D., Olita, S.: The vibrations induced by surface irregularities in road pavements - a matlab approach. Eur. Transp. Res. Rev. 6(3), 267–275 (2014)
20. Zhao, B., Nagayama, T.: IRI estimation by the frequency domain analysis of vehicle dynamic responses. Procedia Eng. 188, 9–16 (2017)
21. Douangphachanh, V., Oneyama, H.: A Study on The Use of Smartphones Under Realistic Settings to Estimate Road Roughness Condition. https://link.springer.com/article/10.1186/1687-1499-2014-114
22. Staniek, M., Sierpiński, G.: Smart platform for support issues at the first and last mile in the supply chain - the concept of the S-mile project. Sci. J. Sil. Univ. Technol. Ser. Transp. 89, 141–148 (2015)

23. Staniek, M.: Repeatability of Road Pavement Condition Assessment Based on Three-Dimensional Analysis of Linear Accelerations of Vehicles. http://iopscience.iop.org/article/10.1088/1757-899X/356/1/012021/pdf
24. General Director of National Roads and Motorways: Regulation of General Director of National Roads and Motorways on The Diagnosis of Road Pavement Condition and Its Elements. General Director of National Roads and Motorways, Warsaw (2015)

Roundabouts as Elements of Smart and Modern Transport Networks and Systems

Surrogate Measures of Safety at Roundabouts in AIMSUN and VISSIM Environment

Orazio Giuffrè[1], Anna Granà[1(⊠)], Maria Luisa Tumminello[1],
Tullio Giuffrè[2], and Salvatore Trubia[2]

[1] Department of Civil, Environmental, Aerospace and Materials Engineering,
University of Palermo, Palermo, Italy
{orazio.giuffre, anna.grana,
marialuisa.tumminello01}@unipa.it
[2] Faculty of Engineering and Architecture, Kore University, Enna, Italy
{tullio.giuffre, salvatore.trubia}@unikore.it

Abstract. This paper addresses issues on road safety analysis through microscopic traffic simulation models. The Surrogate Safety Assessment Model (SSAM) was applied to read vehicle trajectory files generated by two microsimulators and then calculate surrogate measures of safety. Since safety assessment of any road entity can provide different results based on the microsimulator which is used, the main objective of this research was to estimate the safety performance of three roundabout layouts and compare the conflict events simulated by AIMSUN and VISSIM. The two micro-simulators were used to build the calibrated models of the roundabouts, each fitting the corresponding empirical capacity function. The results provided insights on how to set the SSAM filters in order to have a comparable frequency of conflicts by simulation and examine conditions under which a safety analysis could be independent of the software used.

Keywords: AIMSUN · Microsimulation · Roundabout · SSAM
Surrogate safety measure · VISSIM

1 Introduction

Nowadays modern roundabouts and more recently alternative roundabouts as turbo and flower roundabouts are becoming increasingly attractive to transportation engineers which always try to improve the effectiveness of measures and assessment tools for road safety management.

Traditional approaches to estimate safety performance of road entities need data on road crashes, collection of which requires not only a long observation time, but also further information on the drivers' behavior preceding the crash, the causes behind it, traffic operating conditions at the time of the collision, and so on. Thus, these approaches present a series of shortcomings essentially related to the availability of updated or complete road crash databases. Lacunae in the observed crash data can be overcome by using traffic conflict techniques which analyze the road situations from the aspect of more easily observable traffic conflicts (and other events associated with

© Springer Nature Switzerland AG 2019
E. Macioszek et al. (Eds.): Roundabouts as Safe and Modern Solutions in Transport
Networks and Systems, LNNS 52, pp. 53–64, 2019.
https://doi.org/10.1007/978-3-319-98618-0_5

safety and operations) than crashes, or surrogate measures of safety recently introduced to explore the safety performance of any road facility through simulated vehicle trajectories exported from microscopic simulation models; see e.g. [1]. In this regard, the Surrogate Safety Assessment Model (SSAM) [2] reads trajectories files generated by micro-simulators, calculates the number of potential conflicts, and categorizes these in types based on the directions of colliding path vehicles; for each pair of vehicles involved in the conflict, the SSAM evaluates the surrogate measures of safety. In a recent research, the SSAM was used to predict conflicts at intersections and single-lane roundabouts [1]; the outputs derived from SSAM were compared with the number of the conflicts calculated with crash prediction models and conflicts on-field observed. In turn, the SSAM was also used to evaluate surrogate measures of safety at urban signalized intersections [3]; thus, the conflicts generated by VISSIM [4] and PARA-MICS [5] were compared with crash data. Safety performance of freeway merge areas was also analyzed through the SSAM by using conflicts simulated from VISSIM; the field-measured traffic conflict was used to improve the calibration of the microscopic traffic simulation models and adjust the threshold values in SSAM [6]. Since the assessment of safety performance of road facilities by means of surrogate measures of safety goes through the processing of trajectory files generated by a micro-simulator, it is clear that this safety analysis will strongly depend on microscopic traffic simulation models. Indeed, outputs of micro-simulation, i.e. the trajectory files, not only depend on the choices made by the analyst in the building step of the simulation model, but could also depend on the micro-simulator which is used.

Based on this premise, the goal of the paper was to measure the safety performance of three different schemes of roundabout (i.e. single-lane, double-lane and turbo roundabouts) through surrogate measures of safety. For each roundabout under examination, a comparison was performed based on the trajectory files derived from AIMSUN [7] and VISSIM [4]. Thus, these traffic microsimulation models were used to build the calibrated models of the roundabouts, each fitting the corresponding empirical capacity function based on a meta-analytic estimation of the critical and follow-up headways as performed by Giuffrè et al. [8]. Furthermore, in order to explore the implications of various traffic volume distributions on the safety performance of the selected roundabouts, different traffic flow scenarios were simulated. For each roundabout, origin-destination matrices were then developed by using an iterative process so as to ensure a pre-fixed saturation ratio at each entry; thus, the simulated vehicle trajectories exported from AIMSUN and VISSIM were used to perform a conflict analysis through the Surrogate Safety Assessment Model (SSAM). The SSAM results processed from the two traffic microsimulation models provided a very high number of potential conflicts especially when SSAM default filters were used. Since the outputs from the two software also resulted strongly different, it was deemed necessary to set iteratively some SSAM filters, in order to render the outputs from AIMSUN and VISSIM comparable.

The articulation of the paper is as follows: Sect. 2 describes the calibration of the microscopic traffic models here used. Section 3 explains the assessment of safety performance through surrogate measures of safety for the roundabout case studies, and shows the results in terms of surrogate measures of safety provided by using the SSAM software. Section 4 concludes the paper.

2 Calibration of Microscopic Traffic Simulation Models

In order to compare the SSAM results in terms of frequencies of potential conflicts simulated with AIMSUN and VISSIM, the two microscopic simulation models were calibrated by using the same input data. The calibration process in AIMSUN and VISSIM was conducted based on a sensitivity analysis which identified for each software the calibration parameters with the best performance on simulation outputs of entry capacity. And then, there was the comparison between the model output of capacity at steady state as simulated by AIMSUN and VISSIM for each roundabout and an empirical capacity function based on the Hagring model [9].

The main geometric characteristics of the roundabouts selected as case studies are shown in Table 1. Note that the design characteristics of the single-lane roundabout and the double-lane roundabout meet general roundabout classification as geometric design standards and guidelines propose around the world; see e.g. [10, 11]. In turn, the turbo roundabout geometric design for the selected case study was chosen consistent to the turbo design proposed by Giuffrè et al. [12].

Table 1. Geometric characteristics of three roundabout layout selected as case studies.

Roundabout	Outer diameter [m]	Circulatory roadway width [m]	Entry lane width [m]	Exit lane width [m]
Single-lane	39.00	7.00	3.75	4.50
Double-lane	41.00	9.00	3.50	4.50
Turbo	40.00	4.50 (inside lane width) 4.20 (outside lane width)	3.50	4.50

The network models of the roundabout case studies were built in AIMSUN and VISSIM; see Fig. 1 for the case of single-lane roundabout models. As is well-known,

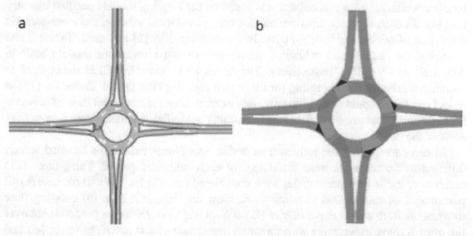

Fig. 1. The single-lane roundabout models built in AIMSUN (a) and VISSIM (b)

AIMSUN and VISSIM work with two microscopic traffic simulation models including car following and lane change logic; each of them has many parameters that affect vehicle interactions and thus the simulation results; for further details see [4, 7]. Once the three roundabouts were represented, O/D matrices were assigned from all entries with due consideration to the directions of turn; thus, a circulating flow (facing the subject entry) from 0 to 1400 veh/h was simulated. A saturated condition was imposed at each entry lane; the corresponding maximum number of vehicles approaching the roundabout gave the entry lane capacity measured through detectors appropriately located. The "car" class was selected as the category of vehicles travelling on the roundabouts. The AIMSUN parameters both for the single-lane roundabout and the double-lane roundabout resulted from the calibration process based on genetic algorithms; see [13, 14]. In both case studies three model parameters were involved: the reaction time, in seconds, that is the time it takes a driver to react to speed changes in the preceding vehicle; the minimum headway, in seconds, that is the minimum distance between the leader vehicle and the follower vehicle; the speed acceptance, that is a user's desired speed, usually interpreted as the 'level of goodness' of drivers of acceptance of speed limits. In turn, the same values of AIMSUN parameters found by Giuffrè et al. [12] for the reaction time and the minimum headway were used for the calibration process of the turbo roundabout.

To calibrate the roundabouts in VISSIM the *car following* model was used; in particular, the *Wiedemann 74* model was selected since it is more suitable for urban traffic. Three parameters were used to calibrate the models: the average standstill distance, that is the average desired distance between stopped cars; the additive part of desired safety distance and the multiplicative part of desired safety distance, which are parameters with major influence on the safety distance and thus affecting the saturation flow rate. Finally, manual calibration of the roundabout models was performed in VISSIM by running the simulation many times, comparing the outputs of simulated capacity with the empirical capacity function, iteratively adjusting the model parameters. To obtain the calibrated models in VISSIM as close as possible to the AIMSUN ones, the model outputs of capacity were compared with the empirical capacity function which for each roundabout was based on the Hagring model specified by entry lane [9]; for each capacity function, one for each roundabout scheme, the meta-analytic estimation of critical and follow-up headways developed by [8] was used. Tables 2 and 3 exhibit the default and calibrated parameters of the roundabout models built in AIMSUN and VISSIM, respectively. The tables also shown the GEH index used to establish a criterion for accepting (or otherwise rejecting) the model. Based on [15], a model can be accepted as significantly able to reproduce the empirical data of capacity if the deviation between simulated and empirical capacities is smaller than 5 in at least 85% of the cases.

In order to explore the influence of traffic operational conditions on road safety, different traffic scenarios were simulated for each calibrated model. Thus, three O/D matrices of traffic flow percentages were considered (see Tables 4, 5 and 6): case (a) all percentages of turning and crossing traffic flow are balanced; case (b) crossing flow movements from and to major entries (East-West and vice-versa) are predominant over the other turning movements with an equal percentage of left and right turns; left and right turning movements from minor to major entries are predominant over through

Table 2. Default and calibrated values of the model parameters in AIMSUN.

Roundabout	Reaction time [s]		Min. headway [s]		Speed acceptance		GEH index [%]	
	Def.	Calibr.	Def.	Calibr.	Def.	Calibr.	Def.	Calibr.
Single-lane	0.80	0.86	0.00	1.58	1.10	1.00	56.25	87.50
Double-lane (right lane)	0.80	0.94	0.00	1.00	1.10	0.95	78.10	96.90
Double-lane (left lane)	0.80	0.95	0.00	1.33	1.10	0.97	78.10	93.80
Turbo[a]	1.35	1.00	0.00	1.70	–	–	85.00[b]	99.00[b]

Note: Def. stands for default parameters; calibr. stands for calibrated parameters; [a]the same values of the model parameters were used for each entry lane; [b]the same GEH indexes were obtained for each entry lane.

Table 3. Default and calibrated values of the model parameters in VISSIM.

Roundabout	ASD [m]		APDSD [m]		MPDSD [m]		GEH index [%]	
	Def.	Calibr.	Def.	Calibr.	Def.	Calibr.	Def.	Calibr.
Single-lane	2.00	5.10	2.00	3.60	3.00	1.80	50.00	87.50
Double-lane (right lane)	2.00	1.80	2.00	3.05	3.00	4.75	50.00	87.50
Double-lane (left lane)	2.00	4.50	2.00	5.00	3.00	5.00	25.00	100.00
Turbo (right lane)	2.00	5.00	2.00	3.10	3.00	1.50	50.00	87.50
Turbo (left lane)	2.00	5.00	2.00	3.10	3.00	1.50	25.00	75.00[a]

Note: ASD: the average standstill distance; APDSD: the additive part of desired safety distance; MPDSD: the multiplicative part of desired safety distance; def. stands for default parameters; calibr. stands for calibrated parameters; [a]for the left entry lane of the turbo roundabout, the GEH index was below 85%, but just a few individual GEH_i value was slightly higher than 5 and the corresponding model could be accepted.

Table 4. Origin/destination matrix of traffic flow percentages for case (a).

Heading level	South	East	North	West
South	0	0.33	0.33	0.33
East	0.33	0	0.33	0.33
North	0.33	0.33	0	0.33
West	0.33	0.33	0.33	0

movements from and to minor entries with a prevalence for left turning movements over the right ones; case (c) similar to case (b), but the percentage of left and right turning movements is now inverted compared to the case (b).

Based on what described by Mauro [11] in Chap. 2, Sect. 2.6, an iterative procedure was then implemented to calculate the total capacity of the three roundabouts and ensure reproduction of a pre-fixed saturation ratio at each entry. Entry capacity was

Table 5. Origin/destination matrix of traffic flow percentages for case (b).

Heading level	South	East	North	West
South	0	0.30	0.05	0.65
East	0.05	0	0.05	0.90
North	0.05	0.65	0	0.30
West	0.05	0.90	0.05	0

Table 6. Origin/destination matrix of traffic flow percentages for case (c).

Heading level	South	East	North	West
South	0	0.65	0.05	0.30
East	0.05	0	0.05	0.90
North	0.05	0.30	0	0.65
West	0.05	0.90	0.05	0

computed by using the well-known relationship among the circulating flow and two parameters depending on the number of circulating lanes and entry lanes; see [16].

It should be noted that for entry capacity calculation at the single-lane roundabout and the double-lane roundabout, the values of parameters to be included in the capacity formula could be uniquely identified for each layout. In turn, for the case study of turbo-roundabout, some assumptions were made based on the operation of each entry lane, i.e. assuming that operation was similar to a single-lane roundabout entry for the right-lane, and a double-lane entry for the left-lane. Thus, all entry lanes of the major road and the right-entry lane of the minor road were treated as the entry lanes of a single-lane roundabout, while the left-entry lane of the minor road was treated as a left entry lane of a double-lane roundabout. Considering a saturation ratio equal to 0.6, for the three roundabouts under examination and based on cases in Tables 4, 5 and 6, nine origin-destination matrices were obtained. In order to extract the trajectory files, traffic demand matrices were inserted into AIMSUN and VISSIM, and for each calibrated model, fifteen replications of simulation, one hour each, were made. From all replications five simulations among the ones that best replicated the O/D matrices were chosen.

3   Comparison of Surrogate Safety Measures from AIMSUN and VISSIM Micro-simulators

As introduced in Sect. 1, the Surrogate Safety Assessment Model (SSAM) software was used to combine micro-simulation and automated conflict analysis and calculate surrogate measures of safety for the roundabouts under examination [17].

Once the trajectory files (named TRJ.files) were separately generated from AIMSUN and VISSIM for each calibrated model of roundabout, the next step was the loading of the trajectory file of each simulation into the SSAM software. Five replications were performed for each case study and the resulting trajectory data generated by the two micro-simulators were analyzed by SSAM; for details on the SSAM performance see e.g. [2, 3]. For each conflict that happened during the simulation the SSAM computed and recorded the following measures identified in the vehicle-trajectory input data: the minimum time-to-collision (TTC), that is the minimum time between two vehicles that will collide with each other if they do not change their respective trajectories; the minimum post-encroachment time (PET), that is the time spent between the end of the crossing vehicle passage and the time at which the through vehicle actually reaches the collision potential point; the initial deceleration rate (DR), that is the initial deceleration rate of the second vehicle, recorded as the instantaneous acceleration rate: if the vehicle brakes, this is the first negative acceleration value observed during the conflict, while if the vehicle does not brake, this is the lowest acceleration value observed during the conflict; the maximum deceleration rate (MaxD), defined as the maximum deceleration of the second vehicle, recorded as the minimum instantaneous acceleration rate observed during the conflict (a negative value indicates deceleration, while a positive value indicates that the vehicle does not decelerate during the conflict); the maximum speed (MaxS), that is the maximum speed of either vehicle throughout the conflict; the maximum speed differential (DeltaS), defined as the magnitude of the difference in vehicle velocities (or trajectories), such that if v_1 and v_2 are the velocity vectors of the first and second vehicles respectively, then DeltaS $= |v_1 - v_2|$.

In order to extract the number of conflicts from SSAM, no threshold filter was initially applied, and the SSAM default filters were kept as the *Filter* folder of SSAM provided; the default values of the maximum TTC of 1.50 s and the maximum PET of 5.00 s were left unchanged when the trajectory files were processed. The conflicts and the relevant surrogate measures of safety recorded in all TRJ.files were then listed in the Conflict folder of SSAM. The results of the SSAM software obtained elaborating the TRJ.files from AIMSUN (i.e. the SSAM-AIMSUN results) and the SSAM results obtained with the TRJ.files from VISSIM (i.e. the SSAM-VISSIM results) were then compared.

Later, in order to better compare the outcome given by the two micro-simulators, the thresholds of the filters were modified. Figure 2 shows the mean values of normalized total conflicts provided by AIMSUN and VISSIM for the examined roundabouts and the traffic situations (called case a, case b and case c) shown in Tables 4, 5 and 6; specifically, the values in Fig. 2 represent the total conflicts by each case study in relation to the total simulated capacity. In the same figure the SSAM-AIMSUN and SSAM-VISSIM are the mean values of total conflicts when the SSAM default filters were maintained unchanged (see Fig. 2a) and the appropriate filters were applied to the surrogate measures of safety as registered by the SSAM (see Fig. 2b). Figure 2a shows significant differences in the mean values of the normalized total conflicts detected by the SSAM which resulted higher for the case studies of double-lane roundabout and turbo roundabout than the single-lane roundabout case study. In addition, especially with TRJ.files coming from AIMSUN, the SSAM found a higher number of conflicts at the turbo roundabout than the double-lane roundabout for the traffic situations in Table 4 (case a) and in Table 5 (case b).

A sensitive analysis was then conducted to identify which filters had an influence on the SSAM results. According to several researches about this issue (see e.g. [18, 19]),

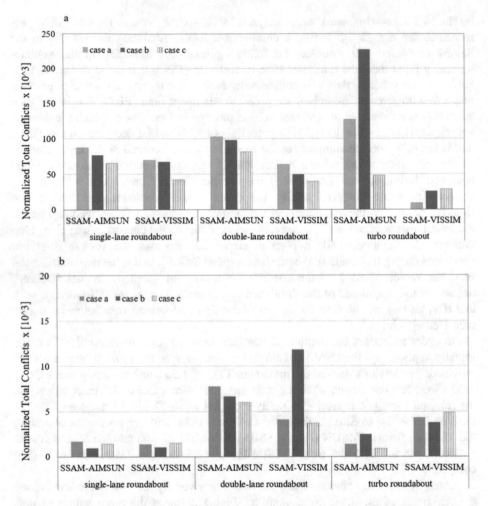

Fig. 2. AIMSUN vs VISSIM total conflicts at roundabout case studies with the SSAM default filters (a) and the SSAM filter-based total conflicts (b). Note: SSAM-AIMSUN and SSAM-VISSIM stand for the mean values of normalized total conflicts obtained with the TRJ.files from AIMSUN and VISSIM, respectively.

the time-to-collision (TTC) and the Post-Encroachment Time (PET) have a higher effect on the outcome of SSAM than other parameters. TTC and PET are both indicators of collision likelihood.

A smaller value of these, during a conflict event, indicates a higher probability of a collision; for instance, a TTC of exactly 0 is a collision by definition, while the PET has the same interpretation of safety as TTC, although possibly not the same magnitude of impact. The Max Speed, defined as the maximum speed of either vehicle throughout the conflict, was identified as another parameter that affects the SSAM results; see e.g. [18]. Several studies agree that conflicts with a TTC value of less than 1.50 s indicate a high probability of collision and therefore consider this value as the maximum threshold [19]. By contrast, there are few references about the critical PET threshold, but in any case, it is clear that a low value of this parameter indicates a higher probability of a crash; moreover, by definition, the value of PET should be greater than the TTC [2]. Thus, after several trials, for all cases studies the TTC and PET maximum thresholds were set of 1.50 s and 2.50 s, respectively. A more restrictive value for PET was applied for the double-lane roundabout: a maximum value of PET of 1.90 was set for conflicts processed with the VISSIM trajectory files. This maximum value of PET is nothing other than the one recorded by SSAM with the AIMSUN trajectory files. In turn, a minimum threshold value of the *Max Speed* was set equal to 1.00 and 1.18 for the single-lane roundabout and turbo roundabout, respectively. No filter of the *Max Speed* was set for the double-lane roundabout. Note that if SSAM registered a minimum value of TTC and PET equal a zero, these thresholds were increased up to 0.10. Indeed, zero values of TTC and PET represent a mere processing error and for this reason it is necessary to delete them; see e.g. [18].

Note that a filter around the area of interest was applied since VISSIM identified conflicts that were very far from the intersection area, i.e. several conflicts were also identified on the legs of the roundabout, upstream of the intersection. Thus, the comparison - about the number of conflicts detected by the SSAM with the TRJ input files coming from AIMSUN and VISSIM - only provided conflicts that fell within a 30-m radius of each roundabout entry. As one can see in Fig. 2, by applying the afore-mentioned filters, a good fit was obtained for the frequency of conflicts derived from AIMSUN and VISSIM. Indeed, the percentage difference of total conflicts generated by the two micro-simulators did not exceed, in any traffic scenario, the percentage of 39%. The student's t-test was used to compare the surrogate measures of safety as recorded by SSAM. The t-test calculated the probability of the difference of the average values of total conflicts for each roundabout under the different traffic scenarios in Tables 4, 5 and 6. It is well-known that the null hypothesis (H0) indicates that no

Table 7. VISSIM vs AIMSUN t-test results at roundabout case studies.

Layout	AIMSUN		VISSIM		t-value	T-student $\alpha = 0.05$		T-student $\alpha = 0.01$	
	Aver.[a]	Var	Aver.[a]	Var		$T_{critical}$	Significant	$T_{critical}$	Significant
Single-lane									
Case a	3.40	2.30	2.80	0.20	0.85	2.31	NO	3.36	NO
Case b	1.80	0.20	2.00	1.00	0.40	2.31	NO	3.36	NO
Case c	3.00	2.00	3.20	0.70	0.30	2.31	NO	3.36	NO
Double-lane									
Case a	19.60	13.30	23.80	0.20	2.56	2.31	YES	3.36	NO
Case b	16.20	1.70	22.60	0.30	10.12	2.31	YES	3.36	YES
Case c	16.00	10.00	19.60	12.80	1.68	2.31	NO	3.36	NO
Turbo									
Case a	3.40	2.80	2.60	3.80	0.70	2.31	NO	3.36	NO
Case b	6.20	1.20	6.00	11.00	0.13	2.31	NO	3.36	NO
Case c	2.20	0.70	3.00	3.50	0.90	2.31	NO	3.36	NO

[a]Average stands for the mean value of total conflicts in the replications

significant difference results between the means of the two samples. Table 7 shows the results of the t-test. The t-test for all the traffic cases in Tables 4, 5 and 6 at single-lane roundabout and turbo roundabout shows that the difference between the means of the two samples (i.e. the SSAM-VISSIM and SSAM-AIMSUN results) are due to the chance. Since the t-test resulted significant for some cases at the double-lane round-about, one can say that there is a significant difference between the means of the two samples and more research is needed.

It should be noted that results in Fig. 2 also show that operational conditions as well as the roundabout layouts have a significant influence on safety at roundabouts.

4 Conclusions

This paper addressed issues on safety analysis through microscopic traffic simulation models based on surrogate measures of safety.

Safety performance of three roundabout schemes through surrogate measures of safety was explored. Based on the simulated vehicle trajectories exported from AIM-SUN and VISSIM, a conflict analysis through the SSAM was performed. Since a safety analysis of any road facility through surrogate measures of safety can be conducted by a lot of common micro-simulators like AIMSUN, VISSIM, PARAMICS and so on, setting filters in SSAM is strictly necessary in order to have similar output data, in term of number of conflicts. A good correlation between the simulated conflicts by AIM-SUN and VISSIM was obtained after iteratively adjusting some SSAM parameters, especially setting lower values of the Time-To-Collision and Post Encroachment Time than default ones and also eliminating those conflicts that provided a zero value. This result is in part confirmed by the t-test, in order to compare the SSAM-VISSIM and SSAM-AIMSUN results for the examined roundabouts under different traffic condition. Indeed, the t-test resulted no significant for the single-lane and turbo roundabouts; on the contrary, this test for the double-lane roundabout had turned out significant, especially for a significance level α of 0.05. Finally, concerning to implications that different traffic scenarios could have on safety performance at roundabouts, it is possible to state that a different flow distribution at roundabouts provides a different number of conflicts; thus, for the same roundabout scheme exists a traffic scenario that provides less potential crashes than another traffic scenario.

The comparison between the surrogate measures of safety based on the simulated trajectories derived from AIMSUN and VISSIM provided insights on how to set the SSAM filters in order to have a comparable frequency of conflicts as the statistical test also confirmed. The results are sufficiently encouraging to suggest that this line of research should continue, with a view to that use of simulated conflicts is a viable, promising approach for estimating the roundabout safety performance. Moreover, the outcome of this research activity could represent the starting point to address, in the near future, issues associated with the development of safety prediction models for roundabouts based on surrogate measures of safety.

This could create a broader framework when the benefit–cost analysis method, used for most public works and transportation projects, should be applied to aid decision about the best suitable solution for a (selected) location based on the type of problem susceptible to correction by the roundabout alternative. Automated road safety analysis through reliable safety evaluation tools based on surrogate measures of safety can result very useful to deliver timely safety estimates to match the safety-affecting progress in vehicles and intelligent infrastructures.

References

1. Vasconcelos, L., Leto, L., Seco, A.M., Silva, B.S.: Validation of surrogate safety assessment model for assessment of intersection safety. Transp. Res. Rec. **2432**, 1–9 (2014)
2. Gettman, D., Pu, L., Sayed, T., Shelby, S.: Surrogate Safety Assessment Model and Validation: Final Report. Technical report FHWA-HRT-08-051. Federal Highway Administration Research and Technology. https://www.fhwa.dot.gov/publications/research/safety/08051/index.cfm
3. Essa, M., Sayed, T.: Transferibility of calibration microsimulation model parameters for safety assessment using simulated conflicts. Accid. Anal. Prev. **84**, 41–53 (2015)
4. PTV Planung Transport Verkehr AG: PTV VISSIM User Manual. PTV Planung Transport Verkehr AG, Karlsruhe (2012)
5. Quadstone PARAMICS. http://www.paaramics-online.com/
6. Chen, P., Zeng, W., Yu, G., Wang, Y.: Surrogate safety analysis of pedestrian vehicle conflict at intersections using unmanned aerial vehicle videos. J. Adv. Transp. **2017**, 1–12 (2017)
7. Aimsun: Aimsun Dynamic Simulator User Manual. Transport Simulation System Version 8. Aimsun, Barcelona (2011)
8. Giuffrè, O., Granà, A., Tumminello, M.L.: Gap-acceptance parameters for roundabouts: a systematic review. Eur. Transp. Res. Rev. **8**(1), 1–20 (2016)
9. Hagring, O.: A Further Generalization of Tanner's Formula. Transp. Res. Part B: Methodol. **32**(6), 423–429 (1998)
10. Tollazzi, T.: Alternative Types of Roundabouts: An Informational Guide. Springer, New York (2015)
11. Mauro, R.: Calculation of Roundabouts: Capacity, Waiting Phenomena and Reliability. Springer, Heidelberg (2010)
12. Giuffrè, O., Granà, A., Marino, S., Galatioto, F.: Microsimulation-based passenger car equivalents for heavy vehicles driving turbo-roundabouts. Transport **31**(2), 295–303 (2016)
13. Giuffrè, O., Granà, A., Tumminello, M.L., Sferlazza, A.: Estimation of passenger car equivalents for single-lane roundabouts using a microsimulation-based procedure. Expert Syst. Appl. **79**, 333–347 (2017)
14. Giuffrè, O., Granà, A., Tumminello, M.L., Sferlazza, A.: Capacity-based calculation of passenger car equivalents using traffic simulation at double-lane roundabouts. Simul. Model. Pract. Theory **81**, 11–30 (2018)
15. Barceló, J.: Fundamentals of Traffic Simulation. Springer, London (2010)
16. Brilon, W., Wu, N., Bondzio, L.: Unsignalized Intersections in Germany - A State of the Art 1997. In: Proceedings of the 3rd International Symposium on Intersections Without Traffic Signals, pp. 61–70. University of Idhao, Idhao (1997)
17. Tarko, A., Davis, G., Saunier, N., Sayed, T., Washington, S.: Surrogate Measures of Safety. White Paper. http://citeseerx.ist.psu.edu/viewdoc/download?doi=10.1.1.148.6770&rep=repl&type=pdf

18. Saleem, T., Persaud, B., Shalaby, A., Ariza, A.: Can microsimulation be used to estimate intersection safety? Case studies using VISSIM and paramics. Transp. Res. Rec. **2432**, 142–148 (2014)
19. Giuffrè, T., Trubia, S., Canale, A., Persaud, B.: Using microsimulation to evaluate safety and operational implications of newer roundabout layouts for European road networks. Sustainability **9**(11), 1–13 (2017)

Modeling of Traffic Conditions at the Circular Junction in the City of Hlohovec

Jan Palúch, Kristián Čulík, and Alica Kalašová[✉]

The Faculty of Operation and Economics of Transport and Communications,
University of Žilina, Žilina, Slovakia
{jan.paluch,kristian.culik,
alica.kalasova}@fpedas.uniza.sk

Abstract. At present, it is possible to see an increase in car traffic on roads in EU. This trend is caused by still growing demand for transport. This is resulting in a gradual increase in road traffic intensity, congestions and traffic accidents. Therefore, it is necessary to use current road network and build a new, which satisfies the present and future requirements for transport in area. Despite the large investments in the development of infrastructure, the traffic situation is getting worse. The most problematic places in the traffic infrastructure are junctions. In the recent past the circular junctions were highly preferred, but are they also good at increasing traffic? In our article, we analyze the highly loaded circular junction in the city of Hlohovec. We have suggested its solutions with microsimulation.

Keywords: Circular junctions · Calculations of communication capacity
Modeling of circular junctions · Traffic model

1 Introduction

In recent years, due to rising standard of living and systematic economic advantage of car transport, it is possible to see an increase in individual car transport and current traffic network is no longer suitable. Transport collapses, traffic congestions, accidents and other negative impacts on environment are gradually becoming a normal part of life. Therefore, more communications and interchanges are built. Despite the large investments in the development of infrastructure, the traffic situation is getting worse. The intersection is a point in the communication network where flows of vehicles meet, connect, disconnect or cross. In other words, the junction is a place where ground communications intersect or associate from top view and they are connected together [1].

Therefore the junction must have sufficient ability to release all these traffic flows. Otherwise, the vehicles could stay in junction or before the intersection. When we are deciding about type, shape, location and configuration of intersection, the security and continuity of the flows must be assured as best as possible.

With increasing number of vehicles on roads it is necessary to ensure, that its crossing through the junction is continuous and also safe. With appropriate software, it is possible to simulate different types of junctions. The simulation outputs can be further processed and compared with technical conditions and with the real situation.

© Springer Nature Switzerland AG 2019
E. Macioszek et al. (Eds.): Roundabouts as Safe and Modern Solutions in Transport
Networks and Systems, LNNS 52, pp. 65–76, 2019.
https://doi.org/10.1007/978-3-319-98618-0_6

Simulation allows trying different types of intersections in short period of time. Then we can compare each design and we can choose optimal variant for area and intensity of road traffic. Before the final choice of suitable type of junction it is also important to analyze its position in relation to junction roads, transport significance in area, deficiencies of junction and its dimensions.

With this approach, it is possible to ensure continue and safe crossing through junction without significant delays. The output should be the optimal variant of junction that takes into account all its specifics.

Nowadays, in many cities and towns, there are built a lot of small circular junctions. They often replace current junctions with only traffic signs or junctions with traffic lights. Their goal it to better aesthetically fit into the center of the city. However, this often leads to increased congestion problems.

2   Analysis of the Solved Circular Junction

The city of Hlohovec is located in Western Slovakia, in Trnava self-governing region. Its location is very important for transportation, because it connects the surroundings cities such as Nitra, Piešťany and others with capital city Bratislava. The road number II/513 passes through the city and this road is one of the main transport links between Nitra and Bratislava. This road splits the city into two parts. The Hlohovec District has over 45 000 inhabitants and an area of 267 km^2 [2]. On the Fig. 1 is the city of Hlohovec with marked position of the solved small circular junction.

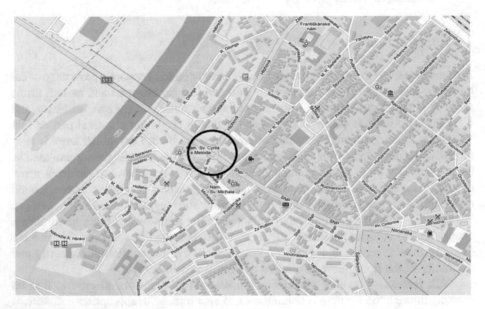

Fig. 1. City of Hlohovec with marked position of the small circular junction km (Source: [3])

At the beginning of the city Hlohovec in the direction from D1 there is a small circular junction. Its location is very important. Various transport directions meet on it. The first mentioned entrance is from road number II/513, which is also connected to the highway D1. Also entrance number 3 to the junction is on this road in direction of Nitra. Entrance number 2 to the junction is from the city center and adjoining settlement. The last north entrance to the small circular junction is the road number II/507 in the direction from Piešťany. The location of inputs is shown in Fig. 2. At this time, there are frequent traffic jams, which sometimes reach a length up to 2 km.

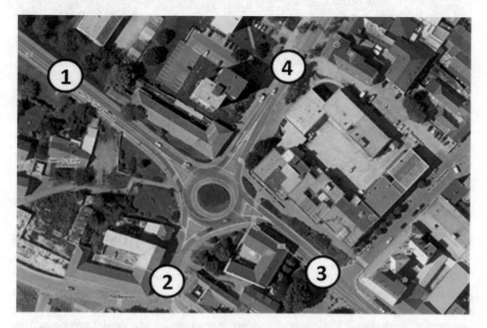

Fig. 2. Entrances to the small circular junction marked with numbers (Source: [3])

The small circular junction has an outside diameter of 38 m. This type of junction is inappropriate for the current traffic load. It is necessary to look for another type. Larger performance than the small circular junction has a traffic-light controlled junction that can give vehicles from one direction better than circular junction. By changing the type of junction, it is possible to improve the traffic situation on the road. Besides that it is also possible to reduce emissions which are produced by cars, because the time spent in the city center will be reduced, too.

3 Evaluation of Traffic Survey on a Solved Junction

At this intersection, a traffic survey was conducted on September 21, 2017 on Thursday. The weather was variable and during the day there were slight precipitations. The survey was conducted using a camera placed on a public lighting column. The view of the junction area from the camera device is shown in the following figure. At the bottom right of Fig. 3 there is a time track allowing accurate evaluation of the traffic load of the small circular junction at individual time intervals of 15 min.

Fig. 3. The view of the junction area from the camera device placed on a public lighting column

From the data, which was collected during traffic survey, it was possible found a peak hour on a small circular junction. The peak hour is in time from 6:45 am to 7:45 am and during this time 2.617 vehicles crossed this junction. Finding their direction was very difficult. For this purpose, the video camera record was used. Directions from entrances of junction are shown in Table 1.

Table 1. Matrix of vehicle directions during peak hour.

	1	2	3	4	SUM
1	–	132	434	255	821
2	246	–		160	631
3	494	54	–	75	623
4	386	49	107	–	542
SUM	1126	235	766	490	2617

All values are numbers of vehicles.

The most loaded entrance of small circular junction was entrance number 1 in the direction from highway D1. Its values during the peak hour reached up to 821 vehicles.

The most loaded exit was also number 1. Its load is up to 1126 vehicles during the peak hour. For better illustration of vehicle flows from individual entrances, see Fig. 4.

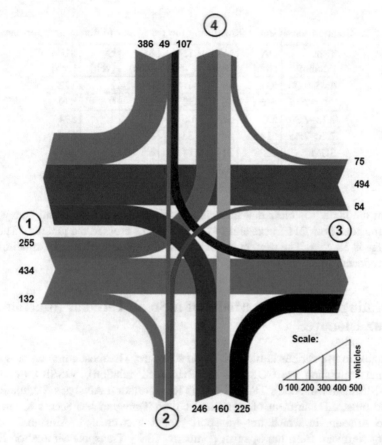

Fig. 4. Load scheme during peak hour

During the traffic survey, observers wrote down five categories of vehicles. These categories were took from Slovak technical conditions 102 - Calculation of the road capacity. Technical conditions 102 also show the conversion factors from real vehicles to unit vehicles. For categories of vehicles were used these abbreviations [1]:

- OA - passenger car,
- NA - truck,
- TNA - heavy truck,
- M - motorbike,
- C - bicycle.

The largest parts in the composition of the traffic flow at the small circular junction during the peak hour were passenger cars. Their number was 2258 vehicles, which is up to 86.28% of the total load during the peak hour. The category of vehicles with the smallest number was category C - only 2 cyclist crossed the junction during peak hour. In the Table 2 is total composition of traffic flow.

Table 2. Composition of traffic flow during the peak hour (different transport means).

Time [hh:mm]	OA [veh]	NA [veh]	TNA [veh]	M [veh]	C [veh]	SUM [veh]
6:45–7:00	537	49	41	0	1	572
7:00–7:15	546	56	32	1	0	579
7:15–7:30	557	56	36	2	1	594
7:30–7:45	618	50	32	2	0	639
SUM	2258	211	141	5	2	2617
Percentage	86.28%	8.06%	5.39%	0.19%	0.08%	–

From the table it is clear that there is a high number of trucks and heavy trucks. During the peak hour 211 trucks and 141 heavy trucks crossed the junction. Their total percentage is 13.45%. The percentage of cyclist and motorcyclist is so small that they are not reflected in simulation.

4   Creating a Transport Model of a Small Circular Junction Near Hlohovec

It is possible to use various traffic simulation software. The most common programmes for virtual simulation are CORSIM (CORridor SIMulation), VISSIM (Verkehr In Städten-SIMulationsmodell), TRANSIMS (TRansportation ANalysis SIMulation System) and others. Department of Road and Urban Transport has access to simulation software Aimsun, in which our transport model was created. Aimsun is a traffic modelling software from the Spanish company TSS - Transport Simulation Systems based in Barcelona. It can perform macroscopic, mesoscopic and microscopic simulation. In Aimsun, it is also possible to model different transport networks: urban networks, freeways, highways, bypasses and their combinations. Aimsun simulation provides a number of outputs that are divided into several groups: statistics for whole network, statistics for group of segments, statistics for sections and digressions, track statistics, statistics for the source/destination matrix and statistics for public passenger transport. Outputs are generated for each group. These outputs are intensity, density, average speed, section speed, travel time, time of stay, stop time, number of stops, total distance travelled, total travel time, fuel consumption and amount of produced emissions [4].

4.1   Creating of Map Layer

First step of creating a map layer was importing the communication network of specific area of road network. The communication network was imported from the database

Open Street Map, which has kept the correct scale. The images of the monitored area were then imported.

4.2   Creating of Road Network of Specific Area

With the Aimsun functions, all entrances to the juncture were created and adjusted. Then the width of lanes, the length of segments and maximum speeds were modified.

4.3   Creating Traffic Model

Two traffic models were designed, but they are different the following conditions:

- the first recorded the current situation at a small circular junction,
- the second planned to rebuild the current small circular junction to a light-crossed junction.

Fig. 5. Designed transport model of circular junction

The traffic model of traffic junction reflects the actual situation, but creating of the light-crossed junction is more complicated. Due to the high load from the different directions it was suitable to make three connecting links between the input and output 1–2, 2–3, and 3–4. However, the buildings around the junction do not allow a connecting between the input and output 4–1. Created traffic models on the map layer on the scale are shown in Figs. 5 and 6.

Fig. 6. Designed transport model with traffic-light controlled junction

In the picture no. 6 is the junction area marked with a yellow color. The vehicles cannot stop on this marked area, so they do not jam the junction. Another important step for the proper operation of the transport model is the creation of a light signal plan. Signal plan has cycle with duration of 80 s, 2 s offset, and 3 s yellow light. The entire signal plan is shown in the Fig. 7.

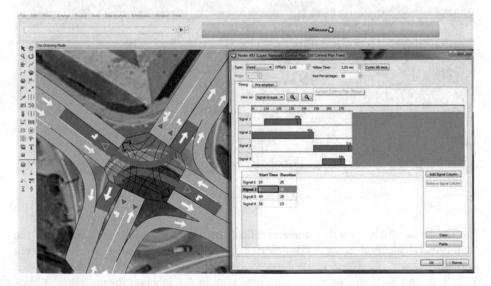

Fig. 7. Designed light signal plan

5 Microsimulation of Traffic Models

For each traffic model were performed 40 simulations. From these simulations it was created average. The running of the simulation of traffic-light controlled junction is in the Fig. 8. In this picture it is possible to see an active signal plan, which give green light to vehicles in rotation and allows for better distribution of traffic over time.

Fig. 8. Running of microsimulation of traffic-light controlled junction

The most important indicators recorded were delay time, stop time, number of stops, travel time, speed, intensity and density. All indicators were analyzed and then they were compared them with the current state. Simulations were performed for two hours, because it is necessary to minimize the influence of starting vehicles on the transport network. The following table shows all decreased values for each category of vehicles.

From the Table 3 is clear, that individual parameters have significantly decreased. The most noticeable decline was recorded at the stop time - the decrease was up 53.32%. Another significant decline was recorded at the delay time and the number of stops, which fell by over 52%. As a result of these decreases, vehicles are losing less time in traffic congestion and have lower fuel consumption. They also produce less emission, carbon dioxide, carbon monoxide, oxides of nitrogen and others [6, 7].

Table 3. Composition of traffic flow during the peak hour (traffic flow parameters).

Parameter	Current status	Traffic light	Unit	Change [%]
Travel time	273.31	163.25	sec/km	−40.27
Stop time	178.73	83.43	sec/km	−53.32
Delay time	204.29	97.33	sec/km	−52.36
Density	36.03	20.10	veh/km	−44.21
Number of stop	0.36	0.17	#/veh/km	−52.78

The Table 4 shows an increase of recorded values of individual parameters for all vehicle categories. These results are due to the improvements of the previous indicators shown in Table 3. They are direct results of improving the traffic situation at the junction [8].

Table 4. Simulation results with increased values of monitored parameters.

Parameter	Current status	Traffic light	Unit	Change [%]
Speed	20.76	30.47	sec/km	+46.77
Traffic flow	2465.5	2603	sec/km	+5.28

Speed has increased by 46.7%, which means increase of almost 10 km/h. Intensity increased by 5.28%. This is due to the fact that there was no blockage of the entrances, the intersection was able to manage higher load and the program was able to put more vehicles on the transport network [5] (Fig. 9).

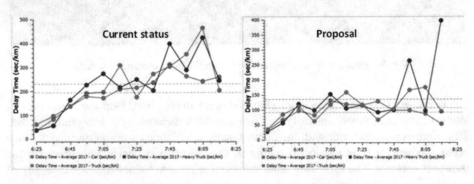

Fig. 9. Delay time progress according to vehicle categories

Figure 10 shows a decrease in delay time. During last stages of simulations the delay time was increased for heavy trucks, but the overall average is much lower.

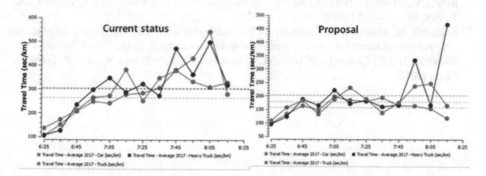

Fig. 10. Travel time progress according to vehicle categories

Travel time is one of the most important indicators. Total average for each category of vehicles is much lower than in the current situation on small circular junction. This indicator has increased during last stages as in the time of delay. See Fig. 10.

6   Conclusion

The current situation at the small circular junction in Hlohovec is critical, which was confirmed by the creation of a transport model and subsequent simulations. With the new design of the reconstruction from a small circular junction to a light-crossed junction, it is possible to reduce values of individual parameter, such as delay time, stop time and travel time, on average over 52%. On the other hand, the speed has increased by 46.7%. This increase is also due to an increase in traffic intensity, because there will be no traffic jams. Our proposal can contribute to a better traffic situation at junction. It can reduce time losses and emissions from vehicles passing through the city of Hlohovec.

References

1. Ministry of Transport and Construction of the Slovak Republic: Technical Conditions 102, Calculation of Road Communications Capacity. Ministry of Transport and Construction of the Slovak Republic, Slovak Republic (2015)
2. Municipality of Hlohovec: Data Provided by Municipality of Hlohovec. Municipality of Hlohovec, Hlohovec (2018)
3. Maps. https://sk.mapy.cz/zakladni?x=17.8079583&y=48.4321355&z=15&base=ophoto
4. Aimsun. https://www.aimsun.com/aimsun-next/editions/
5. Ondruš, J., Černický, Ľ.: Usage of polcam device for parameter monitoring and traffic flow modelling. Commun.: Sci. Lett. Univ. Žilina **18**(2), 118–123 (2016)

6. Konečný, V., Petro, F., Berežný, R.: Calculation of emissions from transport services and their use for the internalisation of external costs in road transport. Perner's Contacts **11**, 68–82 (2016)
7. Rievaj, V., Synák, F.: Does electric car produce emissions. Sci. J. Silesian Univ. Technol. Ser. Transp. **94**, 187–197 (2017)
8. Kalašová, A., Mikulski, J., Kubíková, S.: The impact of intelligent transport systems on an accident rate of the chosen part of road communication network in the Slovak Republic. In: Mikulski, J. (ed.) Challenge of Transport Telematics. CCIS, vol. 640, pp. 47–58. Springer, Cham (2016)

Wireless Electric Vehicles Charging in the Area of Roundabouts

Krzysztof Krawiec[(✉)] and Łukasz Wierzbicki

Faculty of Transport, Silesian University of Technology, Katowice, Poland
{krzysztof.krawiec, lukasz.wierzbicki}@polsl.pl

Abstract. One of the main barriers to the development of electromobility is the inadequate network of vehicle charging points. On the other hand, despite the considerable costs of this technology, the development of technology for wireless charging of electric vehicles in motion is being observed. The article proposes the use of a roundabout area for wireless charging of electric vehicles using the fact that vehicles in the area of roundabouts are moving at low speed. The article presents three variants of the location of inductive charging devices in the area of roundabouts, having regard to the advantages and disadvantages of each solution.

Keywords: Roundabout · Electric vehicle · Inductive charging
Electromobility

1 Introduction

The main purpose of changing the road traffic organization from conventional intersections to the roundabouts is to calm the traffic, increase the traffic safety level and reduce the cost of maintaining the intersection in comparison to the intersections equipped with traffic lights [1–4]. What is characteristic for vehicles approaching and passing through the roundabouts, is the low vehicle speed. If roundabouts aimed to calm the traffic are located in the areas of high traffic volume, there will be queues at the entries.

Calculation of queue lengths and waiting times can be calculated in many ways. Methodology for determining the capacity of roundabouts is presented and discussed in many literature sources and can be found inter alia in Akcelik [5] and Mauro [6]. The calculation of capacity and other traffic parameters is also the subject of consideration in Highway Capacity Manual [7]. With the use of these methods not only the capacity of traffic lanes of conventional roundabouts entries can be determined, but also at turbo roundabouts [8].

As a rule, a formation of vehicle queues at the roundabout's entries is a negative effect, but one can use it for charging electric vehicles. Currently, the number of electric vehicles in use is not yet significant (2 million electric vehicles in 2016), although it is constantly growing. Deployment scenarios for the stock of electric cars to 2030, presented in [9] indicates that electric car stock will up to 20 million by 2020 and up to 40 million by 2025.

© Springer Nature Switzerland AG 2019
E. Macioszek et al. (Eds.): Roundabouts as Safe and Modern Solutions in Transport
Networks and Systems, LNNS 52, pp. 77–86, 2019.
https://doi.org/10.1007/978-3-319-98618-0_7

One of the main barriers to the development of electromobility is the inadequate network of vehicle charging points. What is more, further development of electromobility understood as the number of electric vehicles in operation is strongly dependent on the development of a user-friendly charging network.

However, before we go into a detailed analysis of the idea, let us have a glance at the charging systems for electric vehicles.

2   Electric Vehicles Charging Systems

Electric vehicles and charging infrastructure have three major advantages [10].

The first one is the fact that the electric technology is quite simple and can be realized in ours homes. Electrical systems do not require a complex chain of energy distribution. This system already exists and we use it every day by turn on the source of light, computer, radio or refrigerator. Almost every home can be turned into a charging station of electric cars. However, a design of the charging infrastructure is the basis for the development [11–13].

Secondly, electricity can be transported in two directions. The drive system can convert stored electrical energy into motion but also the kinetic energy of the vehicle back into energy storage through regenerative breaking. Guziński et al. suggested that electric vehicles can be used as energy storages that could balance the electric consumption during energy peaks and pits [14].

Construction of electric vehicles is the last advantage. Electric motors are simpler and have more efficient mechanism than combustion engines. Electric vehicles' torque characteristics are much more suitable to the demand curve of a car. What is more, there is no need for a complex multi-ratio transmission and one or two gear ratios are sufficient [10].

Charging systems for electric vehicles are mainly categorized by the rate at which the batteries are charged. Charging time is one of the most important features of an electric car for the buyer. However, this value depends on many parameters, like for example, the way the battery is depleted, how much energy it holds, the type of battery and the type of charging station. For these reasons, the charging time can range from 15 min up to 10 h or more.

We can distinguish three main types of electric vehicle charging. The first one is charging through electrical conduction. This is a simple method using the connection of electrical wires through the appropriate port to the vehicle. Currently, many types of plugs and sockets are used for both direct (DC) and alternating current (AC). So called "combo" technology is growing strongly that might be a standard for electric vehicles, allowing both DC and AC charging in the same standard of plug [10, 15]. A variation of this method, used mostly in public transport, is charging with the use of a pantograph mounted on the roof of an electric bus.

Charging with alternating current can be conducted in slow charging more or in fast charging mode, depending on current used. Depending on the manufacturer, system use ne or three phase installations. Simple and domestic systems use standardized regular household sockets not exceeding 16 A and not exceeding 250 V AC single-phase or 480 V AC three-phase. It normally takes 6–8 h to fully charge a battery. In the case of more developed charging systems, equipped with electric charging controllers, the charging can take place at different power levels, so it can be either slow or fast. A three-phase system with a 400 V and 63 A current allows charging the battery in time not exceeding one hour. However, this charging systems are more complicated and expensive. Moreover, they require the modification of electrical installation if they are to be a home power station for a car in the garage [10, 15].

For the fast charging we use a direct current charging mode; the supply voltage is 400 V and the top current is 200 A. These parameters of current are quite high, thus they require safety attention and the correct use of appropriate connectors and cables. DC allows a 80% charging of the battery in under 20–35 min [16]. Currently, two main systems are competing with each other: the Japanese CHAdeMO and American Tesla Supercharger. The CHAdeMO system are more worldwide and universal but Tesla's system are more faster and spectacular [10, 15].

Secondly, we can replace empty batteries by charged batteries. The battery swap is the service with a one through an infrastructure similar to a service station. The process is fast and simple. Some battery exchange station are fully automated and it is no need for the driver to leave the car. This is a practical and effective way of bringing the charging time of electric vehicles and the process is fast like refueling in petrol vehicles (about 2–5 min) [10].

A wireless method is the last type of charging. This charging method is based on the inductive phenomenon relies on two coupled coils with magnetic coupling between them. The primary coil is connected with the electric grid and the second one is connected with the battery charging system on board of the vehicle [10, 17, 18]. Fast inductive charging technologies allow the exchange of high power quantities (>20 kW) between an electric vehicle and the electrical grid in contactless way [19]. The wireless charging is the most user-friendly. There is no need for the driver to be worried about handling power cords, thus avoiding the electrocution risk, and can park the car in proper spaces, so that the charging operation automatically starts.

The vehicles can be charged by mainly three technical options: static wireless charging, quasi-dynamic or dynamic wireless charging [10, 20]. The static type of charging happens when the vehicle is stationary and nobody stays inside it, e.g. in the case of a parked car. In the quasi-dynamic, the charging occurs when the electric vehicle is stationary, but someone stays inside it, e.g. in the case of a cab at the traffic light intersections. The dynamic wireless charging happens when the vehicle is moving, e.g. in the case of a car driving on the road. The most attractive advantage of inductive charging is represented by the quasi-dynamic and dynamic types. The disadvantage of dynamic type wireless charging is necessary of building new infrastructure e.g. roadway lanes or creeks with charging option. However, the benefits of recharging a vehicle's battery at critical traffic spots can be significant. Dynamic wireless charging technology can protect the city's road network from traffic jams caused by overestimation of the electric vehicle's range. It should be remembered that many electric cars cannot boast of a large number of kilometers driven on a single long-during charge. The dynamic recharging systems, if widely applied on road network, can allow users to drive in unlimited range without stopping to recharge [21, 22].

Deflorio et al. presented a method for analyzing the traffic and electric performance of dynamic wireless charging systems for two types of electric vehicle: a light-van for freight distribution and a city car [23]. Authors proved that low-length charging area (10 m) could be powered up the battery even on movable vehicle (less than 20 km/h). For this reasons, roundabouts can be the main element of the wireless dynamic charging infrastructure.

3   Variants of Electric Vehicles Charging in the Area of Roundabouts

In order to optimally use the traffic characteristics of vehicles in the area of round-abouts, one should consider the variants of the areas on which the induction charging devices should be located. There are three options for wireless charging vehicles in the area of roundabouts:

- wireless electric vehicle charging at the entries of roundabouts,
- wireless electric vehicle charging in the roundabout apron,
- wireless electric vehicle charging both at the entries and in the roundabout apron.

In the following sub-sections, the various variants of vehicle loading in the area of roundabouts will be described, along with the pictorial drawings and an indication of the advantages and disadvantages of such solutions.

3.1   Wireless Electric Vehicle Charging at the Entries of Roundabouts

Installing chargers at the entries is the first solution that comes to mind as in the case of heavy traffic, these are the areas where the queues are being created. The length of the inductive charging zone should, in principle, be as long as possible, however, it should also depend on the length of the queue to the roundabouts. The length of the queues at the entries of the roundabout depends directly on the traffic observed on these entries. The capacity of roundabouts strongly depends on whether it is a single-lane roundabout or multi-lane one. On the other hand, multi-lane roundabouts are built in places with significantly higher traffic. This means that there may also be queues of vehicles of considerable length on them.

The length of the queue of vehicles approaching the roundabouts can be calculated on the basis of the large literature quoted above. On this basis, it is possible to determine the length of the wireless charging area on which induction charging will be enabled. It should be noted that this length is a dependent variable, which in the article was marked as n. The concept of charging devices located at the entries of the roundabout is presented in Fig. 1. The length of the roadway at the entries of the roundabout is denoted by n.

In the case of uneven traffic load, it may happen that the differences between the queue lengths at individual inlets will be very significant. These differences should be taken into account when designing charging devices only on those entries, which are significantly loaded with road traffic because there is no justification for doing so on entries. This situation is shown in Fig. 1(a), where at a single-lane roundabout, only two entries are equipped with charging devices.

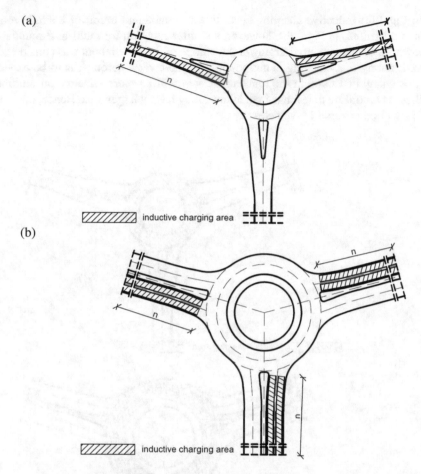

Fig. 1. (a) Inductive charging areas at the entries of single-lane roundabout (b) Inductive charging areas at the entries of multi-lane roundabout

In turn, in Fig. 1b, even traffic intensity was assumed on all entries. In this situation, electric vehicle charging with the use of the wireless method is possible on both lanes while approaching to the roundabout.

3.2 Wireless Electric Vehicle Charging in the Roundabout Apron

Charging electric vehicles with the use of inductive charging technology is possible not only on the entries of the roundabouts, as shown above, but also on the roundabout shields. Unlike roundabouts' entries, where variations in the speed of vehicles depending on the traffic occur, vehicles drive through the roundabout shields at a similar low speed. Hence, it is possible to charge electric vehicles for a small, but generally constant time.

In Fig. 2(a) inductive charging areas in the roundabout apron of a single round-about is presented. In Fig. 2(b), however, a similar situation for multi-lane roundabout is shown. In both cases, no charging at the entries of the roundabout was considered. It is worth noting that charging is not possible in the central apron. It is to be expected that this charging variant in the roundabout area will be more effective on multiband families, as according to technical guidelines, they have a larger area. Hence, the longer section will be covered by vehicles.

(a)

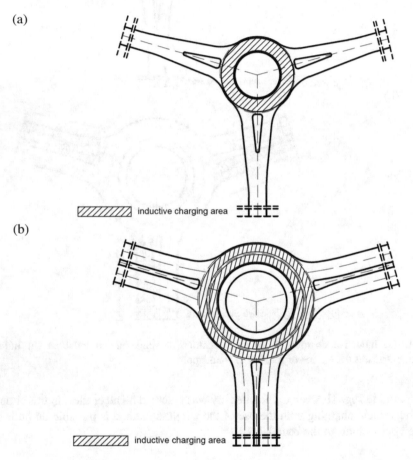

(b)

Fig. 2. (a) Inductive charging areas in the roundabout apron of single-lane roundabout (b) Inductive charging areas in the roundabout apron of multi-lane roundabout

The charging variant on the roundabout has a major disadvantage: it is necessary to adapt the charging infrastructure to the shape of the roundabout (circle, oval), which may be problematic for manufacturers of chargers and thus increase costs of invest-ment. In combination with the significant costs of induction charging, this drawback may be difficult to overcome.

3.3 Wireless Electric Vehicle Charging Both at the Entries and in the Roundabout Apron

By the combination of the two charging methods in the area of the roundabouts discussed in the previous two subsections, we come to the third one: charging both at the entries and in the roundabout apron. This solution is shown in Fig. 3, both for single-lane and multi-lane roundabouts.

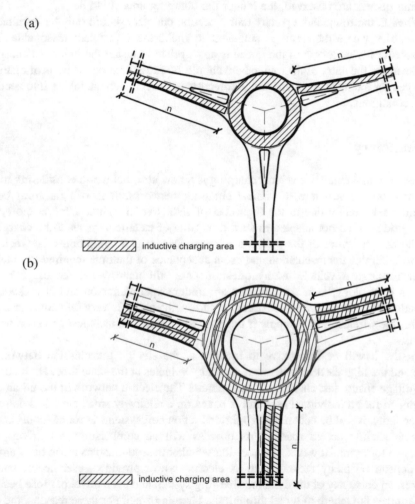

Fig. 3. (a) Inductive charging areas at the entries and in the roundabout apron of single-lane roundabout (b) Inductive charging areas at the entries and in the roundabout apron of multi-lane roundabout

The main advantage of this variant is the maximization of the surface on which it is possible to load vehicles. It also follows that the vehicles will be in the inductive charging area for the longest time, and as a result, they can recharge more. On the other hand, the investment costs will also be the largest, which results from the loading surface on the inlets and the and the need to adjust the shape of the inductive charging area to the shape of the roundabout shield.

To use this variant of vehicle loading, it is necessary to analyse the traffic conditions at the roundabout to calculate the length of the queues at the entries. Similarly to the first variant (charging at the entries only), the length of the vehicle charging area is n and depends on the traffic observed at the roundabout - the more traffic and the resulting queues are observed, the longer the charging area could be.

When at the roundabout patchy traffic occurs, chargers should only be installed in these entries where the traffic is sufficient to make the investment reasonable. The assessment of what length of the queue is appropriate to install the inductive charging devices under the entry of the roundabout should be carried out on the basis of financial resources for the investment and traffic forecasts at the roundabout, taking into account traffic conditions.

4 Summary

The use of roundabouts for vehicle charging is a new idea, but worth considering in the discourse on the structure of vehicle charging networks. It is still far from being implemented, mainly due to the high cost of inductive charging. What is more, the power grid is often not adapted to such a number of electric vehicles to be charged.

The design layout of the dynamic wireless system is a compromise between the need to minimize the installation and users acceptance of the time required to obtain a proper recharge a vehicle. Many questions are still unanswered for the dynamic wireless technology, as more solutions are under experimentation and at present, an estimation of costs related to construction and maintenance is very difficult. For a few reasons, the wireless power supply is not easy to introduce in technical and economical aspects.

Firstly, it will be imperative to modernise the electric transmission network to withstand the high electricity demand for many vehicles at the same time. High current and voltage ensure fast charging of the vehicle. The electric network in the urban area is intended for an individual user and is based on a relatively small electrical demand.

Secondly, it will be necessary to build induction coil systems in roundabouts and/or on them leading access roads. This process will be cumbersome for drivers and expensive for users. However, the question whether the roundabouts are to be equipped may depend on many factors, such as electric power provided, user needs, battery features, an efficiency of the dynamic charging system and even travel purpose because it determines the length of travel through the charging zone. For these reasons, the cost of building the wireless charging system will depend on the variant of the location of the inductive zones on the roundabouts and the ambient conditions. However, the analysis of these issues would require specific studies, which are out of the scope of the presented work.

Thirdly, the area of the roundabout should ensure a small distance between the coils in the vehicle and the ground. The closer the coils are, the higher the efficiency of the wireless charging system is. For this reason, the roadway in the loading area should be flat, without inspection chambers, road humps etc.

Fourthly, the selection of roundabouts will be important. Induction charging zones should be in places with slow-moving vehicles. However, it should be passable enough so that it would not be necessary to quickly rebuild it.

References

1. Nicolae, F., Golgot, C.: Traffic optimization in urban area - roundabout versus lights case studies. In: Chiru, A., Ispas, N. (eds.) CONAT 2016 International Congress of Automotive and Transport Engineering, pp. 694–701. Springer, Cham (2017)
2. Macioszek, E.: The comparison of models for follow-up headway at roundabouts. In: Macioszek, E., Sierpiński, G. (eds.) Recent Advances in Traffic Engineering for Transport Networks and Systems. LNNS, vol. 21, pp. 16–26. Springer, Cham (2018)
3. Macioszek, E.: Analysis of significance of differences between psychotechnical parameters for drivers at the entries to one-lane and turbo roundabouts in Poland. In: Sierpiński, G. (ed.) Intelligent Transport Systems and Travel Behaviour. AISC, vol. 505, pp. 149–161. Springer, Cham (2017)
4. Macioszek, E.: The comparison of models for critical headways estimation at roundabouts. In: Macioszek, E., Sierpiński, G. (eds.) Contemporary Challenges of Transport Systems and Traffic Engineering. LNNS, vol. 2, pp. 205–219. Springer, Cham (2017)
5. Akcelik, R., Chung, E., Besley, M.: Roundabouts: Capacity and Performance Analysis. ARRB Group Limited, Vermont South (1998). Research Report
6. Mauro, R.: Calculation of Roundabouts: Capacity Waiting Phenomena and Reliability. Springer-Verlag, Heidelberg (2010)
7. Transportation Research Board: Highway capacity manual 2010. Transportation Research Board of the National Academy of Science, Washington (2010)
8. Macioszek, E.: The application of HCM in the determination of capacity of traffic lanes at turbo roundabout entries. Transp. Prob. **11**(3), 77–89 (2016)
9. International Energy Agency: Clean energy ministerial, electric vehicles initiative: global ev outlook 2017. Technical report. https://www.iea.org/publications.freepublications/publication/GlobalEVOutlook2017.pdf
10. Longo, M., Zaninelli, D., Viola, F., Romano, P., Miceli, R., Caruso, M., Pellitteri, F.: Recharge stations: a review. In: 2016 Eleventh International Conference on Ecological Vehicles and Renewable Energies, pp. 1–8. IEEE Press, New York (2016)
11. Micari, S., Polimeni, A., Napoli, G., Andaloro, L., Antonucci, V.: Electric vehicle charging infrastructure planning in a road network. Renew. Sustain. Energy Rev. **80**, 98–108 (2017)
12. Brandstätter, G., Kahr, M., Leitner, M.: Determining optimal locations for charging stations of electric car-sharing systems under stochastic demand. Transp. Res. Part B **104**, 17–35 (2017)
13. He, F., Yin, Y., Zhou, J.: Deploying public charging stations for electric vehicles on urban road networks. Transp. Res. Part C **60**, 227–240 (2015)
14. Guziński, J., Adamowicz, M., Kamiński, J.: Infrastructure for charging electric vehicles. Autom. Electr. Disrupt. **1**, 74–82 (2014)
15. Martínez-Lao, J., Montoya, F.G., Montoya, M.G.: Electric vehicles in Spain: an overview of charging systems. Renew. Sustain. Energy Rev. **77**, 970–983 (2017)
16. Aziza, M., Odaa, T.: Simultaneous quick-charging system for electric vehicle. Energy Procedia **142**, 1811–1816 (2017)

17. García-Vazquez, C.A., Llorens-Iborra, F., Fernandez-Ramírez, L.M., Sanchez-Sainz, H., Jurado, F.: Comparative study of dynamic wireless charging of electric vehicles in motorway, highway and urban stretches. Energy **137**, 42–57 (2017)
18. Bi, Z., Kan, T., Mi, C.C., Zhang, Y., Zhao, Z., Keoleian, G.A.: A review of wireless power transfer for electric vehicles: prospects to enhance sustainable mobility. Appl. Energy **179**, 413–425 (2016)
19. Karakitsios, I., Karfopoulos, E., Hatziargyriou, N.: Impact of dynamic and static fast inductive charging of electric vehicles on the distribution network. Electric Power Syst. Res. **140**, 107–115 (2016)
20. Jeong, S., Jang, Y.J., Kum, D.: Economic analysis of the dynamic charging electric vehicle. IEEE Trans. Power Electron. **30**(11), 6368–6377 (2015)
21. Liu, H., Wang, D.Z.W.: Locating multiple types of charging facilities for battery electric vehicles. Transp. Res. Part B **103**, 30–55 (2017)
22. Fuller, M.: Wireless charging in California: range, recharge, and vehicle electrification. Transp. Res. Part C **67**, 343–356 (2016)
23. Deflorio, F., Guglielmi, P., Pinna, I., Castello, L., Marfull, S.: Modeling and analysis of wireless "charge while driving" operations for fully electric vehicles. Transp. Res. Procedia **5**, 161–174 (2015)

Data Collection, Data Analysis
and Development of Model and Methods
for Roundabouts

Gap Acceptance Cycles for Modelling Roundabout Capacity and Performance

Rahmi Akçelik[(⊠)]

SIDRA SOLUTIONS, Melbourne, Australia
rahmi.akcelik@sidrasolutions.com

Abstract. Gap-acceptance theory has been used widely for estimation of capacity at unsignalised roundabouts and two-way sign-controlled intersections. This paper discusses the use of the gap-acceptance method beyond modelling capacity. The author has developed gap acceptance capacity and performance models by signal analogy, including the estimation of delay, queue length and stop rates for roundabouts and other unsignalised intersections. These models have been implemented in the SIDRA INTERSECTION software which has been in extensive use in traffic engineering practice. This paper will describe the basic method that uses gap acceptance cycles for modelling performance measures with a focus on the modelling of queue length at roundabouts. A simple single-lane roundabout example is given to explain important aspects of modelling the queue length.

Keywords: Intersection · Roundabout · Sign control · Signals
Queue · Back of queue · Cycle-average queue · Delay · Stop rate
Gap acceptance · Critical gap · Follow-up headway

1 Introduction

Gap-acceptance theory has been used widely for estimation of capacity at unsignalised intersections including roundabouts and sign-controlled intersections that operate by the give-way (yield) rule. There are numerous capacity models in the literature based on this approach [1]. This paper discusses the use of gap-acceptance theory beyond modelling capacity. The author has developed gap acceptance capacity and performance models for unsignalised intersections by signal analogy, including the estimation of delay, queue length and stop rates, which in turn helps with the estimation of fuel consumption and emissions for unsignalised intersections. These models have been implemented in the SIDRA INTERSECTION software which has been in extensive use in traffic engineering practice. Detailed information about these models is available in a large number of papers and reports and are available for download from [2].

This paper will describe the basic method that uses gap acceptance cycles with a focus on the modelling of queue length at roundabouts with some reference to capacity and delay modelling as well. The method is applicable to two-way sign (give-way and stop) control as well. A simple single-lane roundabout example is given to explain important aspects of modelling the queue length.

© Springer Nature Switzerland AG 2019
E. Macioszek et al. (Eds.): Roundabouts as Safe and Modern Solutions in Transport
Networks and Systems, LNNS 52, pp. 89–98, 2019.
https://doi.org/10.1007/978-3-319-98618-0_8

2   Modelling Roundabout Capacity by Gap Acceptance Cycles

The main purpose of this paper is to the discuss modelling of two types of queue length that can be observed at roundabouts, namely the back of queue and the cycle-average queue, demonstrate the significant difference between these two queue types, and emphasise the importance of using the back of queue rather than the cycle-average queue in single intersection and network modelling.

Figure 1 depicts the modelling of gap acceptance cycles and its application to the modelling of capacity, delay, queue length, and other performance measures at unsignalised intersections.

A gap acceptance cycle consists of a *blocked period* and an *unblocked period*, i.e. vehicles waiting due to lack of an acceptable gap and then departing when an acceptable gap occurs, similar to a signal cycle that consists of a red period and a green period.

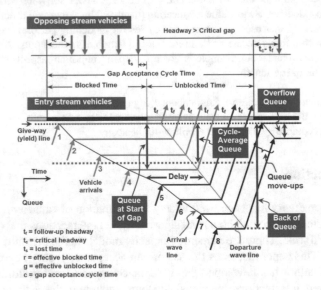

Fig. 1. An oversaturated gap acceptance cycle showing different queue length types

The capacity determined by this method can be expressed in a simple general form as follows [1, 3, 4]:

$$Q = s \cdot u \tag{1}$$

where:

Q - capacity [veh/h],
u - the proportion of time when the vehicles can depart from the queue,
s - saturation flow rate [veh/h].

Equation (1) is the same as the capacity equation for signalized intersections where u is the green time ratio. For gap-acceptance processes at roundabouts and sign-controlled intersections, u is the *unblocked time ratio*, i.e. the ratio of the *unblocked time* (when gaps in the opposing stream are acceptable) to the *gap acceptance cycle time* (sum of *blocked time* and *unblocked time*).

Saturation flow rate (s) corresponds to a *queue discharge headway* representing the minimum headway between vehicles that is achieved while they are departing from the queue:

$$h_s = \frac{3600}{s} \tag{2}$$

where:

h_s - queue discharge (saturation) headway [s],
s - saturation flow rate [veh/h].

The gap-acceptance method uses the *follow-up headway* (t_f) as the queue discharge (saturation) headway, $t_f = h_s$. Therefore:

$$s = \frac{3600}{t_f} \tag{3}$$

where:

s - saturation flow rate [veh/h],
t_f - follow-up headway as a queue discharge (saturation) headway [s].

As seen in Fig. 2, the follow-up headway corresponds to a saturation flow rate which is the maximum gap-acceptance capacity that can be achieved when the opposing flow is close to zero.

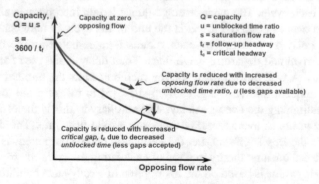

Fig. 2. The gap acceptance capacity

Thus, the saturation flow rate for a gap-acceptance process is the maximum gap-acceptance capacity that can be achieved when the opposing flow is close to zero. In Fig. 2, this is seen as the *y intercept* of the capacity curve. The capacity is reduced from this value with increased opposing flow rates due to the decreased values of *unblocked time ratio.*

In capacity models based on gap-acceptance modelling, while the follow-up headway determines the capacity value at low opposing flow rates directly, the critical gap parameter affects the *unblocked time ratio* (u) with lower values of u resulting from larger values of critical gap (hence lower capacity) for a given opposing flow rate (circulating flow rate for roundabouts). This is also depicted in Fig. 2.

3   Back of Queue and Cycle Average Queue

The *back of queue* has been used commonly for modelling signalised intersection performance. On the other hand, the general literature and various guidelines as well as traffic theory text books present only the *cycle-average queue* based on traditional gap acceptance and queuing theory models as relevant to unsignalised intersections. This discrepancy continues to exist in the signalised and unsignalised intersection chapters of US Highway Capacity Manual Edition 6 [5].

Figure 1 shows an oversaturated gap acceptance cycle with various queue length types. These are the *back of queue, cycle-average queue, queue at start of gap* and *overflow queue.* Figure 1 also indicates the relationship between *queue move-ups* (multiple stops) and the *overflow queue.*

In Fig. 1, idealised vehicle trajectories are shown. Vehicles 1 to 4 arrive during a blocked interval (no acceptable gaps). Thus, *queue at start of gap* (at start of the unblocked interval) is 4 vehicles. Vehicles 5 to 8 arrive during the unblocked interval, slowing down to join the back of queue. Thus, the *back of queue* in this gap acceptance cycle is 8 vehicles. Vehicles 1 to 6 accept the available gap and depart from the queue (enter the roundabout circulating road). Vehicles 7 and 8 cannot accept the available gap and stop. Thus, they form an *overflow queue* (the gap acceptance cycle is oversaturated).

The *cycle-average queue* is the average value of the number of vehicles in the queue during each cycle. The cycle-average queue length incorporates all queue states including zero queues observed towards the end of the cycle in undersaturated cycles.

In Fig. 1, delay experienced by each vehicle is represented by the horizontal line between the arrival and departure wave lines. Total delay is the area formed by these horizontal lines. A well-known delay survey method counts the number of vehicles in the queue in frequent intervals, e.g. every 5–10 s [5] to measure the total delay, and uses this for estimating the average delay. The corollary to this is the estimation of the *cycle-average queue* as average delay times the arrival flow rate. The delay used for this purpose is the *stopline delay,* i.e. the geometric delay component is not included.

The traditional queuing theory method of calculating the *cycle-average queue* using the average delay value is based on the assumption of steady-state conditions. This may not be reliable when the delay estimates are based on time-dependent queuing theory. In particular, this may result in a mismatch between delay and the *cycle-average queue* for oversaturated conditions if the delay estimate includes the delay experienced by vehicles beyond the analysis period, i.e. the delay experienced until the vehicles (arriving during the analysis period) depart from the queue. This may result in an estimate of the *cycle-average queue* that is larger than the *back of queue*.

The *back of queue* is maximum extent of the queue that occurs once each cycle, usually during the green time at signalised intersections or unblocked time in gap acceptance processes. Zero queue states are not relevant to the back of queue.

The *back of queue* is a more useful performance measure since it is relevant to the design of appropriate queuing space, e.g. for short lane design to avoid queue spillback into adjacent lanes, for phasing design to avoid blockage of upstream intersection lanes in networks situations, and for signal coordination offset design to prevent interruption of platoons by downstream queues. The *back of queue* is used for the prediction of such statistics as the saturated portion of the green period and for modelling short lane capacities.

An interesting aspect of the relation between delay and the *back of queue* is that these performance measures are not necessarily consistent in terms of magnitude. This is reflected in the comparison of the *cycle-average queue* and the *back of queue*. *Low delay* associated with a *long back of queue* as seen in Fig. 3 is a result of a high arrival flow rate, large green time ratio (relatively short red) at signalised intersections or large *unblocked time ratio* (relatively short blocked time) in gap acceptance processes.

The case shown in Fig. 3 corresponds to high capacity and low degree of saturation conditions. For roundabouts and two-way sign control, this case occurs under low circulating/opposing flow and high entry demand flow conditions. In such cases, delay consists of acceleration and deceleration (slow down) delays only, and very small or zero idling (stopped) delays occur. While the large back of queue represents a moving queue formed by a heavy arrival flow, there may be a large proportion of vehicles that are undelayed, and therefore the *cycle-average queue* is usually small in this case.

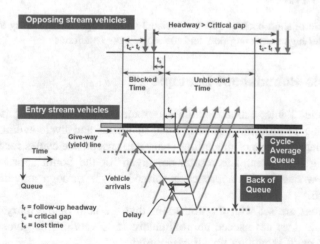

Fig. 3. The case of long back of queue associated with a low average delay at roundabouts: this case occurs under low circulating flow and high entry flow conditions

This case helps to understand why the *back of queue* rather than the *cycle-average queue* should be used for modelling short lanes in intersection modelling and blockage of upstream intersection lanes (queue spillback) in network modelling. This case also has an important role in unbalanced roundabout conditions since the majority of departures with follow-up headways at this entry result in a uniform headway distribution at the next (downstream) entry which leads to low capacity for that approach.

On the other hand, the case of *short back of queue* associated with a *large average delay* as seen in Fig. 4 is a result of a low arrival flow rate and a small green time ratio (relatively long red) at signalised intersections or small *unblocked time ratio* (relatively long blocked time) in gap acceptance processes. This case corresponds to low capacity and high degree of saturation conditions. For roundabouts and two-way sign control, this case occurs under high circulating/opposing flow and low entry flow conditions.

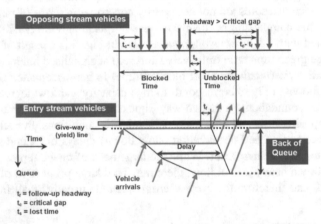

Fig. 4. The case of short back of queue associated with a large average delay at roundabouts: this occurs under high circulating flow and low entry flow conditions

4   A Simple Roundabout Example

The single-lane T-intersection roundabout example shown in Fig. 5 is used to demonstrate the relationship between back of queue and cycle-average queue and present the related aspects of modelling using gap acceptance cycles for varying entry and circulating flow rates. The results are given for the South approach lane. The circulating flow rate for this approach is formed by the through movement from the West approach.

The volumes are set with the constraint that the sum of the entry flow and the circulating flow does not exceed approximately 1500 veh/h. This gives a reasonable range of degrees of saturation for all cases used.

The example is presented using the SIDRA INTERSECTION standard software setup for driving on the right-hand side of the road. To keep the discussion at a basic level, only the average queue length results are given and the percentile queue lengths are not discussed. Using a single-lane roundabout example, complications related to multi-lane roundabouts, e.g. calculation of lane flow rates, effect of unequal circulating flow rates, and so on are excluded.

The analysis results are presented in Figs. 6, 7, 8, 9 and 10. Figure 6 shows the entry capacity as a function of the circulating flow rate for arrival flow rates of $q_a =$ 300, 600 and 1000 veh/h. It is seen that capacities for the three arrival flow rates differ for low circulating flow rates. This is due to the effect of the ratio of entry flow rate to the circulating flow rate (higher values of this ratio give higher capacities in the model). This is an important feature in modelling unbalanced flow conditions.

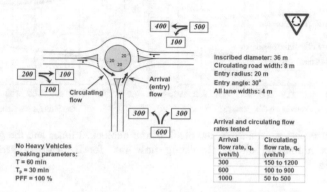

Fig. 5. A single-lane roundabout example to demonstrate the relationship between back of queue and cycle-average queue and the related aspects of modelling

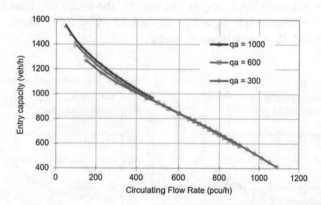

Fig. 6. Entry capacity as a function of the circulating flow rate for arrival flow rates of $q_a = 300$, 600 and 1000 veh/h

Figure 7 shows the gap acceptance parameters (critical gap and follow-up headway), the blocked and unblocked times and the gap acceptance cycle time as a function of the circulating flow rate for the case of arrival flow rate $q_a = 300$ veh/h. It is seen that the critical gap and follow-up headway values are reduced with increased circulating flow rates. This is based on the research at Australian roundabouts [6]. The slight increase in values of these parameters for very low circulating flow rates is related to the model used for the effect of the ratio of the entry flow rate to the circulating flow rate (Medium level of this effect was specified for this example).

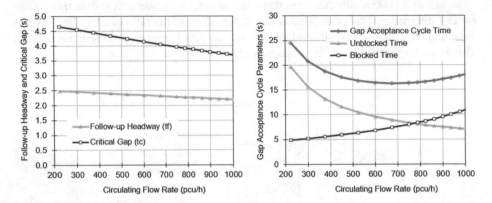

Fig. 7. Gap acceptance parameters, the blocked and unblocked times and the gap acceptance cycle time as a function of the circulating flow rate for the case of arrival flow rate $q_a = 300$ veh/h

Figure 8 shows the *average back of queue* and *cycle-average queue* as a function of the degree of saturation for arrival flow rates of $q_a = 300$ and 1000 veh/h. The correlation of the *average back of queue* and *cycle-average queue* for arrival flow rates of $q_a = 300$, 600 and 1000 veh/h is given in Fig. 9. The linear trendline for these data points gives the relationship $N_b = 1.22\ N_c + 1.82$ ($R^2 = 0.94$) where N_b = average back of queue and N_c = cycle average queue.

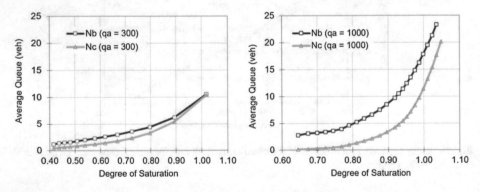

Fig. 8. Average back of queue (N_b) and cycle average queue (N_c) as a function of the degree of saturation for arrival flow rates of $q_a = 300$, 600 and 1000 veh/h

It is seen from Figs. 8 and 9 that the difference between the values of *average back of queue* and *cycle-average queue* increase with increasing arrival flow rate. This is related to the case depicted in Fig. 3.

Average back of queue and stopline delay values used in calculating the cycle-average queue values are given Fig. 10 as a function of the circulating flow rate for arrival flow rates of q_a = 300, 600 and 1000 veh/h.

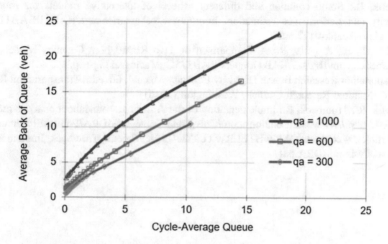

Fig. 9. Comparison of the back of queue and cycle-average queue

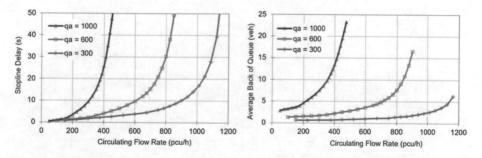

Fig. 10. Stopline delay and average back of queue as a function of the circulating flow rate for arrival flow rates of q_a = 300, 600 and 1000 veh/h

5 Concluding Remarks

Further research is recommended to examine the results given in this paper by means of microsimulation analysis and real-life surveys. The results given here are for a simple case of single-lane roundabout used for the purpose of this paper. The research should consider complications that arise in real-life situations, e.g. the effect of short lanes, variations in various geometric and driver behavior parameters, slip lanes, effect of upstream signals, effect of pedestrians, and so on.

References

1. Akçelik, R.: A review of gap-acceptance capacity models. In: 29th Conference of Australian Institutes of Transport Research, CAITR. University of South Australia, Adelaide. http://eng.monash.edu.au/civil/assets/document/research/centres/its/caitr-home/prevcaitrproceedings/caitr2007/akcelik-caitr2007.pdf
2. SIDRA SOLUTIONS. http://www.sidrasolutions.com/Resources/Articles
3. Akçelik, R.: Some common and different aspects of alternative models for roundabout capacity and performance estimation. http://www.techamerica.com/RAB11/RAB11Papers/RAB1117Akcelik-0127.pdf
4. Macioszek, E., Akçelik, R.: A Comparison of Two Roundabout Capacity Models. http://techamerica.com/RAB17/RAB17papers/RAB175C_MacioszekPaper.pdf
5. Transportation Research Board: Highway Capacity Manual, 6th edn. Transportation Research Board, National Research Council, Washington (2016)
6. Akçelik, R., Troutbeck, R.: Implementation of the Australian roundabout analysis method in SIDRA. http://www.sidrasolutions.com/Cms_Data/Contents/SIDRA/Folders/Resources/Articles/Articles/~contents/QXF24HTELRWTL3XU/1991_Akcelik_Troutbeck_Implemnt_Aus_Rou_analysis-inSIDRA.pdf

Making Compact Two-Lane Roundabouts Effective for Vulnerable Road Users: An Assessment of Transport-Related Externalities

Paulo Fernandes[✉] and Margarida Coelho

Department of Mechanical Engineering, Centre for Mechanical Technology and Automation, University of Aveiro, Campus Universitário de Santiago, Aveiro, Portugal
{paulo.fernandes,margarida.coelho}@ua.pt

Abstract. Compact two-lane roundabouts are increasingly popular. Designing cycle lanes at two-lane roundabouts may not benefit motor vehicles, pedestrians and cyclists simultaneously. This study addresses environmental and operational aspects for accommodating bicycle treatments at compact two-lane roundabouts, namely: (i) sharing bicycles with the motor vehicle lanes; (ii) sharing bicycles with pedestrian pathways; (iii) dedicated bicycle lanes separated from pedestrian paths and motor vehicle lanes. Each scenario was subjected to different traffic, pedestrian and cyclist volumes. Using a microscopic traffic model, the operational performance of the above designs was compared. Then, a microscopic emission methodology based on vehicle-specific power and a semi-dynamic model were used to estimate pollutant emissions and traffic noise, respectively. It was found that cyclists travel time increased with the adoption of separated bicycle lanes since this design led to longer paths. However, average intersection travel time, emissions and noise decreased when compared to other designs.

Keywords: Roundabouts · Cyclists · Pedestrians · Emissions Noise

1 Introduction and Objectives

The modern roundabouts emerged in the United Kingdom (UK) in the 1960s in response to operational and safety problems with old traffic circles built in the beginning of the 20th century [1]. This new intersection treatment included an "off-site priority" rule to govern roundabout operations, namely: (1) entering traffic must give the way to the circulating traffic; (2) vehicles travelling further outside are not privileged in a conflict over the vehicles on the inner lanes. These rules have prompted the rise of modern roundabouts worldwide [1].

Brilon separates roundabouts into six basic categories, according to size, traffic volumes and number of approach, circulating and exit lanes [2], as follows:

- mini-roundabouts have a fully traversable central island, and a diameter between 13 and 23 m. They could carry up to 17 000 vehicles per day (vpd),

The original version of this chapter was revised: The Acknowledgement section has been updated. The correction to this chapter is available at https://doi.org/10.1007/978-3-319-98618-0_16

E. Macioszek et al. (Eds.): Roundabouts as Safe and Modern Solutions in Transport Networks and Systems, LNNS 52, pp. 99–111, 2019.
https://doi.org/10.1007/978-3-319-98618-0_9

- single-lane roundabouts with a diameter between 26 and 35–40 m, and single-lane entries and exits only. The typical daily service is approximately 25 000 vpd,
- compact two-lane roundabouts with inscribed circle diameters vary from 40 to 60 m, 8–10 m lane widths, single or two-lane entries, only single-lane exits, and a maximum capacity of 32 000 vpd,
- conventional multi-lane roundabouts provide two-lane entries and exits, and typical inscribed diameters ranging from 46 to 91 m. These roundabouts can handle between 35 000 and 40 000 vpd,
- turbo-roundabouts contain continuous spiral paths where entry, circulating, and exit lanes are usually separated by raised curbs. They can carry up to 35 000 vpd depending on the arrangements of lanes at the entries and exits,
- signalized roundabouts are used for larger traffic volumes (50 000–60 000 vpd) [2].

Although multi-lane and compact two-lane roundabouts can handle high traffic volumes, these layouts may represent specific problems for most vulnerable road users. This occurs for the following reasons [1]:

- longer crossing distances for pedestrians,
- high pedestrian travel time since pedestrians must assure that all lanes are free of moving traffic,
- drivers are less likely to yield to pedestrians due to the higher speeds,
- higher approaching, circulating and exiting speeds that leads to higher risk propensity of pedestrian or cyclist injury in the event of a collision.

Thus, pedestrian and cyclist treatments present specific design challenges, even to experienced transport planners. It is well known that roundabouts are safer for pedestrians than for cyclists [3, 4] and that roundabout design involves involve the balancing of competing objectives [5]. For example, slower approaching and exiting travelling speeds for increased pedestrian safety, or surfacing treatments for the visually impaired may reduce the available capacity of intersection [6]. Alternatively, designing roundabouts for traffic circulating at similar on-road bicyclist speeds (19 to 32 km/h) speeds improve cyclist's safety, but may represent a risk for pedestrians [1].

The analysis of impedance caused by pedestrian and cyclists at roundabouts is very well documented. With respect to pedestrians, the past research has focused on the analysis of roundabout capacity in the presence of pedestrians [7–9], crosswalk design [6] and location [5, 10], and driving behaviour characteristics [9, 11].

Early studies carried out on cyclists activities along roundabouts have dealt with the analysis of road user behaviour [12, 13], and the testing of different bicycle treatments [14, 15]. Stanek showed that single-lane roundabout delay was higher when more bicycles used roadway, and lower when bike riders shared path with pedestrians [14]. He found that the separation of bike and pedestrian paths showed as the safest solution [14], but the impacts of bicycle treatments on roundabout emissions and noise levels were not included.

Interest also is growing about the relationship between the propensity of yielding to vulnerable users with sharing bicycles with pedestrian pathways and separated bicycle lanes, as studied by Fernandes and Coelho [16]. This research neglected, however, the impacts on traffic noise.

Since good design is not a one-size fits all approach, following a conceptual design methodology may not bring the planned benefits for drivers, pedestrians and cyclists. To date, the difficulty of evaluating different cyclist treatments at compact two-lane roundabouts according to a specific road user criterion is somewhat lacking.

This paper evaluates the impact of several cyclist treatments at a compact two-lane roundabout on traffic, pedestrian and cyclist performance, pollutant emissions and traffic noise.

The novelty purpose of the study is therefore the recognition of the trade-off among traffic-related variables, and the improvement of the compact two-lane roundabout operations according to the road users' needs.

2 Methodology

2.1 Bicycle Design

Two options for accommodating bicycles at two-lane roundabouts are suggested in the "Roundabouts: An Informational Guide" [1]. For confident riders, bicycles travelling in a bicycle lane merge with traffic on the roundabout approach or in a dedicated bike-lane, they traverse roundabouts like vehicles. The use of bicycle lanes within the circulating area of roundabout is not recommended. For concerned riders, bicycles are provided with a ramp to the shared path for bicycles and pedestrians that circles the roundabout [1].

Alternatively, the Dutch roadway design manual provides another option to accommodate bicycles at roundabouts. In this case, bicycles are provided a pathway separated from both motor vehicles and pedestrians [15]. Specifically, bicycles travelling in a bike lane remain in a separate while travelling through the roundabouts. Drivers yield to bicycles and pedestrians at crosswalks, but bicycles yield to pedestrians [15].

2.2 Operations Modelling

2.2.1 Traffic, Pedestrian and Bicycle

To assess the overall operational performance of the compact two-lane roundabout designs, the microscopic traffic model VISSIM 9.0 was used [17]. Link coding was made so that vehicles stated at least one-time step (second-by-second basis) in each link. This assured vehicle dynamic and traffic information for accurate emissions and noise estimates [18].

Figure 1 exhibits the screen captures of the roundabout layout with the proposed bicycle treatments: Typical - sharing bicycles with the motor vehicle lanes or with pedestrian pathways [1]; Protected - dedicated bicycle lanes separated from pedestrian paths and motor vehicle lanes [15]. For both designs, the compact four-leg two-lane roundabout was designed according to the Portuguese guidelines [19]: (a) 50 m inscribed circle diameter; (b) 10 m lane width; and (c) crosswalk located 15 m downstream of the exit junction. Also, approaching vehicles were coded to use outer circulating lane only for right turning.

a)   b)

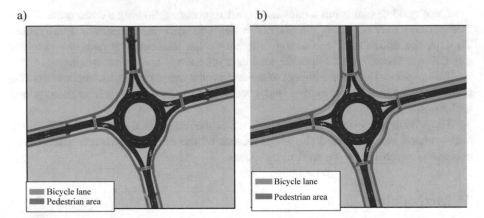

Fig. 1. Proposed bicycle treatments at the compact two-lane roundabout: (a) typical; (b) protected

VISSIM model calibration was made by adjusting the driver behaviour parameters (standstill distance, additive and multiple parts of safety distance, minimum gap time and headway [17]) and examining their effect on roundabout capacity. Equation 1 gives the capacity of a one-lane roundabout entry conflicted by two circulating lanes and subjected to a pedestrian impedance effect, as follows [20]:

$$C_{e,pce} = 1.420 \times e^{\left(-0.85 \times 10^{-3}\right)V_{c,pce}} \times f_{ped} \tag{1}$$

where:

$C_{e,pce}$ - capacity of the entry leg [passenger cars per hour (pc/h)],
$V_{e,pce}$ - conflicting flow (total of both lanes) [pc/h],
f_{ped} - pedestrian impedance to vehicles [-].

Since the capacity of a roundabout approach is influenced by local driving habits, a generalized Siegloch model (Eq. 2) was also used to reflect Portuguese conditions:

$$C_{e,pce} = \frac{3600}{t_f} \times e^{\left(-\left(\frac{t_c - 0.5 \times t_f}{3600}\right)\right)V_{c,pce}} \times f_{ped} \tag{2}$$

where:

$C_{e,pce}$ - capacity of the entry leg [pc/h],
$V_{e,pce}$ - conflicting flow [pc/h],
t_f - critical headway [s],
t_c - follow-up headway [s],
f_{ped} - pedestrian impedance to vehicles [-].

If the number of conflicting pedestrians per hour (p/h) is lower than 101, f_{ped} is computed using Eq. 3 [20]:

$$f_{ped} = 1 - 0.000137 \times N_{ped} \tag{3}$$

where:

f_{ped} - pedestrian impedance to vehicles [-],
N_{ped} - number of conflicting pedestrians per hour [p/h].

The full range of critical and follow-up headway values for Portugal reality are given elsewhere [21]. The calibration procedure considered a demand of 1 550 pc/h on one approach to reach a flow that exceeds compact two-lane roundabout capacity [20]. Thus, traffic volumes at the other approaches were increased to generate a range of conflicting flow rates [20]. For each $V_{e,pce}$, the entering capacity was obtained by using data collection points [17] that record simulated hourly volumes.

The default values were found to be close to the HCM curve for $V_{e,pce} < 150$ pc/h but they did not reflect local driving habits (Fig. 2). A minimum gap time of 3.5 s generated a capacity curve that was closest to the Portuguese adjustment curve ($t_f = 2.3$ s and $t_c = 3.5$ s) [21]. The calibrated parameters intercept (zero conflicting flow) of 1 502 vehicles per hour per lane (vphpl) while the Portuguese adjustment model intercept was 1 490 vphpl. However, the simulated entering capacity was lower than Portuguese model for higher conflicting flows ($V_{e,pce} > 600$ pc/h). This happened because some conflicting cars used outer circulating lane for through turning which reduced the number of available gaps.

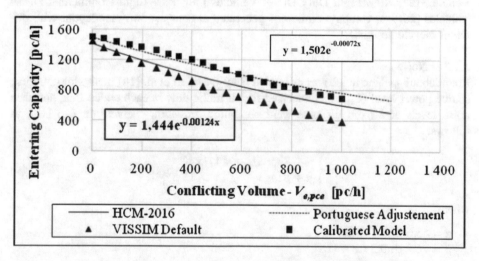

Fig. 2. Model calibration of the VISSIM capacity model

Therefore, the resulting calibration settings were then applied to all four approaches along with the following modelling assumptions:

- warm-up period of 15 min, and data extracted for the following four 60-min interval,
- pedestrian volumes of 50 p/h at the crosswalks,

- no Heavy-Duty Vehicles and Transit Buses traffic,
- the effect of grade was ignored,
- a speed distribution from 14 km/h to 35 km/h for cyclists [22],
- an average pedestrian walking speed value of 1.5 m/s [23],
- free-flow speed from 45–55 km/h for motor vehicles,
- average speed from 22–27 km/h for motor vehicles at the roundabout approaches and at the roundabout circulating area,
- average speed from 30–35 km/h for motor vehicles after leaving roundabout.

2.2.2 Emissions

VISSIM allows computing vehicle activity data (speed, acceleration, and grade) on a second-by-second analysis. Thus, emissions were estimated based on the Vehicle Specific Power (VSP) methodology that represents the sum of the loads resulting from aerodynamic drag, acceleration, road grade and rolling distance effects, all divided by vehicle mass [24].

VSP is widely-recognized as an important driving behaviour parameter and it reflects the changes in vehicle dynamic on emissions at compact two-lane roundabouts [25]. VSP values are categorized into 14 modes, and an emission factor for each mode allows estimating the footprints of carbon dioxide (CO_2), carbon monoxide (CO), nitrogen oxides (NO_X) and hydrocarbon (HC) emissions [24].

This study used a Portuguese fleet composition [26] - 39% Light Duty Gasoline Vehicles [27]; 40% Light Duty Diesel Vehicles [28]; 21% Light Commercial Diesel Vehicles [28] - to compute emissions generated at compact two-lane layout with different bicycle designs.

2.2.3 Noise

Roundabout-specific noise was estimated from Quartieri et al. [18] methodology. First, source power level ($L_{w,i}$) was analysed for all traffic flow in each coded link, and then noise levels at a fixed distance were computed. Equation 4 gives the $L_{w,i}$ [18], as follows:

$$\begin{cases} 82, & if \quad v < 11.5 \ \frac{km}{h} \\ \alpha + \beta \log v, & if \quad v > 11.5 \frac{km}{h} \end{cases} \tag{4}$$

where:

$\alpha = 53.6 \pm 0.3$ dBA,
$\beta = 26.8 \pm 0.2$ dBA [18].

Using the link-specific average speed and volumes (measured directly in the VISSIM), the hourly equivalent noise level can be calculated using Eq. 5 [29]:

$$L_{eq,i} = 10 \log N + \alpha + \beta \log v - 20 \log d - 47.563 \tag{5}$$

where:

$L_{eq,i}$ - link-specific equivalent noise level [dBA],
N - link-specific hourly traffic volume [vph],
v - link-specific average speed [$\frac{km}{h}$],
d - distance between the road axis and the receiver [m] = 7.5 [18].

2.3   Operational and Design Scenarios

The main idea of the evaluation framework was to get quantitative parameters that allowed identifying the merits of the studied bicycle treatments. Rather than obtaining absolute indicators, it is important to understand the circumstances under which a specific bicycle facility may be better than the other for a given road user.

The effects of uniform pedestrian and cyclist growth were explored by varying the number of pedestrians and bike users from 80 to 120 pedestrians per hour (pph) and cyclists per hour (cph) at each crossing. The above demand scenarios were then applied to scenarios with motor vehicle volumes of 1 500 and 2 000 vph, and with the following model assumptions:

- no U-turns were considered,
- traffic distribution was 70% for eastbound and westbound approaches and 30% for northbound and southbound approaches,
- directional splits of 20-60-20 (percentages of right turning, through and left turning movements, respectively) for all road users.

Cyclists may travel through the roundabout as vehicles or they may travel as pedestrians by taking the shared path. To assess the impact of these options on roundabout operations, the percentage of bicycles taking the shared path with cars was set at 0% and 50%, subjected to the following assumptions: (1) bicycles have the same yielding behaviour on roundabout entry as motor vehicles; (2) motor vehicles yield to bicycles in the transition from the bike lane to the roadway; (3) bicycles traveling on the shared path have the right-of-way to pedestrians [14]. Regarding the protected design, motor vehicles were coded to yield to all vulnerable road users, and bicycles were coded to yield to pedestrians [15]. A total of 8 and 4 combinations were tested for the typical and the protected designs, respectively (Table 1).

3   Results

This section compares the impacts for all typical and protected designs. The Typical design with 0% shared path with cars (S1, S4, S7 and S10) was used as the reference for analyses.

3.1   Performance

Table 2 presents the effect of varying vehicle, cyclist and pedestrian demand and cyclist route choice on roundabout performance.

Table 1. Scenarios description.

Scenario	Condition	Vehicles [vph]	Pedestrian [pph]	Bike users [cph]
S1	Typical, 0% path	1 500	80	80
S2	Typical, 50% path	1 500	80	80
S3	Protected	1 500	80	80
S4	Typical, 0% path	1 500	120	120
S5	Typical, 50% path	1 500	120	120
S6	Protected	1 500	120	120
S7	Typical, 0% path	2 000	80	80
S8	Typical, 50% path	2 000	80	80
S9	Protected	2 000	80	80
S10	Typical, 0% path	2 000	120	120
S11	Typical, 50% path	2 000	120	120
S12	Protected	2 000	120	120

Table 2. Variation of traffic performance parameters in relation to the typical design with 0% path.

Scenario	Total Stops	Roundabout Travel Time [s/veh]	Cars Travel Time [s/veh]	Travel Time for Bicycles [s/veh]	Pedestrian Travel Time [cph]
S1	256	66.3	36.4	75.0	191.9
S2	15 %	-1 %	3 %	-9 %	0 %
S3	-14 %	-2 %	-5 %	-11 %	-1 %
S4	342	78.4	37.6	81.5	197.5
S5	40 %	-1 %	6 %	-9 %	-1 %
S6	-13 %	-5 %	-6 %	-14 %	-2 %
S7	766	73.4	39.4	75.5	192.2
S8	10 %	-1 %	4 %	-5 %	0 %
S9	-13 %	-3 %	-2 %	-9 %	-1 %
S10	1 063	82.5	41.8	82.1	197.6
S11	40 %	3 %	9 %	-8 %	-1 %
S12	-13 %	-2 %	-3 %	-14 %	-2 %

Note: Average values using 20 runs [30]; Shadow cells indicate that the difference between scenarios and typical design with 0 % path output measure was statistically significant (p-value < 0.05).

The trade-off between cyclist and cars travel time was clearly observed between designs. When 50% of bicycles shit to using the roadway, the average cyclist travel time decreased 9% while cars travel time increased more than 2% for both S2 and S5. The change in bicyclist route choice had slight impacts on resulting average intersection travel time (1%). However, the shift to roadway path worsened roundabout performance under high traffic volumes (S11). On average, it yielded 3% higher travel time compared to S10. This occurred for two reasons: (1) when travelling as a vehicle, bicycles induce low vehicle speeds at the approach area of the roundabout; (2) vehicles tend to have stops due to the higher conflicting flow in the circulating roadway. The results also showed that the pedestrian travel time did not vary among designs (p-value <0.05).

For a given demand of 1 500 vph, S3 and S6 protected designs provided a significant advantage in roundabout performance compared to the S1 and S4 typical designs; its implementation allowed that the number of idling situations and average intersection travel time to be reduced by 13% and 5%, respectively, with S6 conditions. Also, it became more effective in improving roundabout operations when entering demands increases to 2 000 vph. For instance, if each crossing has 120 cph and pph (S12), the decrease in the number of vehicle stops and intersection travel time may be 13% and 2%, respectively, in relation to the S9.

Interestingly, the path route in protected design was longer than the shift to roadway path in 50% but resulting travel times were lower. The explanations for this fact may be in slower-moving pedestrians who obstruct bicyclists along sidewalks and crosswalks. This phenomenon did not occur in the protected design since bicycles maintained their desired speeds through the roundabout influence area even though they yield to pedestrians near crosswalks. These findings are in line with a previous study on a single-lane roundabout [14].

3.2 Pollutant Emissions and Traffic Noise

Figures 3 and 4 depict vehicular emissions and noise generated by traffic by scenario, respectively.

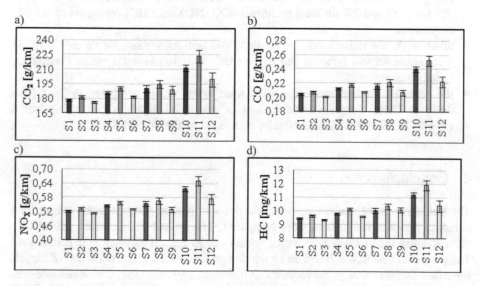

Fig. 3. Emissions levels (with standard error of the mean) per scenario: (a) CO_2 per kilometer; (b) CO per kilometer; (c) NOX per kilometer; (d) HC per kilometer

The following conclusions about the proposed bicycle designs can be drawn:

• as suspected, the emissions per kilometer increased as the demand levels increased mainly in the cases where half of bicycles travelled via the roadway. This point was

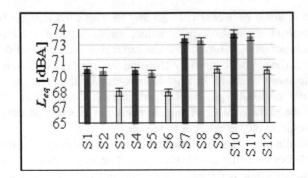

Fig. 4. Noise levels (with standard error of the mean) per scenario

explained by the significant number of stop-and-go and idling situations in the approach area,

- the findings pointed out small differences in emissions (2–3%, depending on the pollutant) between shared path with cars and pedestrian typical designs with a roundabout traffic volume of 1 500 vph (S2/S1 and S5/S4),
- the protected roundabout designs showed as the best emission scenarios, regardless of the demand levels. For instance, if roundabout has 2 000 vph and 120 pph and cph (S12), then the separate bicycle facility through the roundabout could save up to 6% for CO_2 and 7% for local pollutants (CO, NOX and HC) compared to the S10 conditions,
- albeit high, the noise levels ($\sim$70 dBA) did not differ among scenarios (Fig. 4). The reason for this fact was due to the small variation in link-specific average car speeds and volumes,
- the protected designs were associated with lower values compared to the typical designs. There were decreases in Leq of about 2.7 dBA ($\sim$3%) and 3.3 dBA ($\sim$4%) in S9 and S12, respectively, compared to the S6 and S10.

4 Conclusions

The performance, vehicular emissions and noise of typical and protected designs for a compact two-lane roundabout were compared. The typical design either guided bike users to share the roadway with motor vehicles or to share a pathway with pedestrians. The protected design was based on Dutch design guidelines, and it consisted of bicycle separated facilities though the roundabout influence area and located between sidewalk and vehicle roadway.

A microsimulation environment was used to explore changes in traffic, bicycles and pedestrian flow rates and bicycle route choice (0% and 50% bicycles share path with cars). It was found that cyclist travel time was lower when more bicycles used roadway. However, no significant differences in roundabout-specific travel time were observed. For relatively high traffic, bicycle and pedestrian volumes (2 000 vph, 120 cph and pph), this shift between roadway and path routes yielded a range of about 3% on roundabout intersection travel time.

Despite the longer crossing distances for cyclists, the adoption of the protected design resulted in significant travel time reductions for most of the typical design scenarios. The emission and noise benefits of such design was also clear. Under very high demand, the emissions levels decreased from 6% to 7% (depending on the pollutant) while traffic noise decreased approximately 4% compared to the typical designs.

This study contributes to assess several bicycle treatments at a compact two-lane roundabout, and its expected impacts for cars, bike users and pedestrians. This methodology would help local authorities in decision-making process at the intersection level (according to the site-specific conditions and road user needs) in the domain of mobility and emissions.

Further research could be conducted to include the impacts of heavy-duty trucks and transit buses on roundabout-specific operations. The comparison of conflict points among vehicles, pedestrians and bicycles would provide more evidence for when a specific design should be selected over another. These further refinements would be also applied to improve to the HCM methodology for compact two-lane layouts.

Acknowledgments. The authors acknowledge the projects: PTDC/EMS-TRA/0383/2014, that was funded within the project 9471-Reinforcement of RIDTI and funded by FEDER funds; Strategic Project UID-EMS-00481-2013-FCT and CENTRO-01-0145-FEDER-022083; Mobi-Wise project: From mobile sensing to mobility advising (P2020 SAICTPAC/0011/2015), co-financed by COMPETE 2020, Portugal 2020 - Operational Program for Competitiveness and Internationalization (POCI), European Union's ERDF (European Regional Development Fund), and the FCT. This work is financed by ERDF Funds through the Operational Program Competitiveness and Internationalization - COMPETE 2020 and by National Funds through FCT - Foundation for Science and Technology within the scope of the POCI-01-0145-FEDER-16740 project.

References

1. Transportation Research Board: Roundabouts: An Informational Guide. Second Edition. National Cooperative Highway Research Program, Report NCHRP 672. Transportation Research Board, Washington (2010)
2. Brilon, W.: Safety of roundabouts: an international overview. In: TRB 95th Annual Meeting Compendium of Papers, Washington (2016)
3. Sakshaug, L., Laureshyn, A., Svensson, A., Hyden, Ch.: Cyclists in roundabouts—different design solutions. Accid. Anal. Prev. **42**(4), 1338–1351 (2010)
4. Zahabi, S., Strauss, J., Manaugh, K., Miranda-Moreno, L.: Estimating potential effect of speed limits, built environment, and other factors on severity of pedestrian and cyclist injuries in crashes. Transp. Res. Rec. **2247**, 81–90 (2011)
5. Fernandes, P., Fontes, T., Pereira, S.R., Rouphail, N.M., Coelho, M.: Multicriteria assessment of crosswalk location in urban roundabout corridors. Transp. Res. Rec. **2517**, 37–47 (2015)
6. Findley, D.J., Searcy, S.E., Schroeder, B.J.: Investigation of crosswalk design and driver behavior at roundabouts. In: TRB 96th Annual Meeting Compendium of Papers, Washington (2017)

7. Knoop, V.L., Daganzo, C.F.: The effect of pedestrian crossings on traffic flow. In: TRB 96th Annual Meeting Compendium of Papers, Washington (2017)
8. Kang, N., Nakamura, H.: Estimation of roundabout entry capacity that considers conflict with pedestrians. Transp. Res. Rec. **2517**, 61–70 (2015)
9. Schroeder, B., Rouphail, N., Salamati, K., Bugg, Z.: Effect of pedestrian impedance on vehicular capacity at multilane roundabouts with consideration of crossing treatments. Transp. Res. Rec. **2312**, 14–24 (2012)
10. Fernandes, P., Guarnaccia, C., Teixeira, J., Sousa, A., Coelho, M.: Multi-criteria assessment of crosswalk location on a corridor with roundabouts: incorporating a noise related criterion. Transp. Res. Procedia **27**, 460–467 (2017)
11. Salamati, K., Schroeder, B.J., Geruschat, D.R., Rouphail, N.M.: Event-based modeling of driver yielding behavior to pedestrians at two-lane roundabout approaches. Transp. Res. Rec. **2389**, 1–11 (2013)
12. Polders, E., Daniels, S., Casters, W., Brijs, T.: Identifying crash patterns on roundabouts. Traffic Injury Prev. **16**(2), 202–207 (2015)
13. Silvano, A.P., Ma, X., Koutsopoulos, H.N.: When Do Drivers Yield to Cyclists at Unsignalized Roundabouts? Empirical Evidence and Behavioral Analysis. https://www.diva-portal.org/smash/get/diva2:800906/FULLTEXT01.pdf
14. Stanek, D.: Operations and Safety of Separated Bicycle Facilities at Single Lane Roundabouts. https://www.researchgate.net/publication/316884812_Operations_and_Safety_of_Separated_Bicycle_Facilities_at_Single_Lane_Roundabouts
15. CROW: Design Manual for Bicycle Traffic. Dutch Information and Technology Platform. Crow, Netherlands (2017)
16. Fernandes, P., Coelho, M.C.: Pedestrian and cyclists impacts on vehicular capacity and emissions at different turbo-roundabouts layouts. Transp. Res. Procedia **27**, 452–459 (2017)
17. Planung Transport Verkehr: PTV VISSIM 9 User Manual. Planung Transport Verkehr, Karlsruhe (2016)
18. Quartieri, J., Iannone, G., Guarnacci, C.: On the improvement of statistical traffic noise prediction tools. http://www.wseas.us/e-library/conferences/2010/Iasi/AMTA/AMTA-36.pdf
19. Silva, A.B., Seco, Á.M.: Dimensionamento de Rotundas-Disposições Normativas. Instituto de Infra-Estruturas Rodoviárias, Lisbon (2012)
20. Transportation Research Board: Highway Capacity Manual 6th Edition. A Guide for Multimodal Mobility Analysis. Transportation Research Board of the National Academy of Science, Washington (2016)
21. Vasconcelos, A.L., Seco, A.M., Silva, A.B.: Comparison of procedures to estimate critical headways at roundabouts. Promet - Traffic Transp. **25**(1), 43–53 (2013)
22. Micro Simulation of Cyclists in Peak Hour Traffic Guide. http://vision-traffic.ptvgroup.com/en-uk/references/project-references/cowi-cyclist-in-peak-hour-traffic/
23. Chandra, S., Bharti, A.: Speed distribution curves for pedestrians during walking and crossing. Procedia - Soc. Behav. Sci. **104**, 660–667 (2013)
24. United States Environment Protection Agency: Methodology for Developing Modal Emission Rates for EPA's Multi-Scale Motor Vehicle & Equipment Emission System. Report EPA420-R-02-027. Environmental Protection Agency, Ann Arbor (2002)
25. Fernandes, P., Salamati, K., Rouphail, N.M., Coelho, M.: The effect of a roundabout corridor's design on selecting the optimal crosswalk location: a multi-objective impact analysis. Int. J. Sustain. Transp. **11**(3), 206–220 (2017)
26. Emisia. Mission For Environment. http://emisia.com/products/copert-data

27. Anya, A.R., Rouphail, N.M., Frey, H.Ch., Liu, B., Anya, A.: Method and case study for quantifying local emissions impacts of transportation improvement project involving road realignment and conversion to multilane roundabout. In: 92nd Transportation Research Board Annual Meeting, Washington (2013)
28. Coelho, M.C., Frey, H.Ch., Rouphail, N.M., Zhai, H., Pelkmans, L.: Assessing methods for comparing emissions from gasoline and diesel light-duty vehicles based on microscale measurements. Transp. Res. Part D: Transp. Environ. **14**(2), 91–99 (2009)
29. Guarnaccia, C.: Advanced tools for traffic noise modelling and prediction. Trans. Systems **12**, 121–130 (2013)
30. Winnie, D., Christine, B., Serge, P.: Traffic Simulation and Data: Validation Methods and Applications. CRC Press, Taylor & Francis Group, Boca Raton (2014)

Estimating Roundabout Delay Considering Pedestrian Impact

Nan Kang[✉] and Shintaro Terabe

Faculty of Science and Technology, Tokyo University of Science, Noda, Japan
kangnan@rs.tus.ac.jp, terabe@rs.noda.tus.ac.jp

Abstract. Delay at roundabout is commonly estimated based on the entry traffic lane capacity. Pedestrians have significant impact on entry capacity. Entry process becomes complicated when pedestrians are included. This research aims at examining whether the existing delay estimation at roundabouts are available or not after considering pedestrian impacts under the Japanese situation. Microscopic simulation VISSIM and a theoretical regression formula proposed by the author are applied.

Keywords: Roundabout · Delay · Pedestrian impact · Capacity
VISSIM

1 Introduction

Delay estimation plays an important role on operational evaluation. At roundabouts, delay in the existing methods is commonly estimated through entry capacity. Moreover, pedestrians significantly influence the entry capacity. In various guidelines and manuals such as HCM [1], HBS [2] and AUSTROADS [3], pedestrian impact has been considered. However, since space is limited in Japan, some geometric elements such as physical island cannot be installed. This absence indicates that a model which can reflect pedestrian impact and Japanese characteristics is necessary. Thus, the author developed a theoretical regression model considering pedestrian and geometry under the Japanese situations.

On the other hand, entry process becomes complicated when considering pedestrians. It is worth to know whether the existing formula of delay estimation at roundabout is available or not when directly input the entry capacity considering pedestrians. Thus, this research aims at examining the delay estimation at single-lane roundabout (main type in Japan currently) considering pedestrian impact through applying author's model. Since oversaturated situation in Japan is a little difficult to be observed in current state, microscopic simulation VISSIM is applied to provide delay data instead of observation.

© Springer Nature Switzerland AG 2019
E. Macioszek et al. (Eds.): Roundabouts as Safe and Modern Solutions in Transport
Networks and Systems, LNNS 52, pp. 112–123, 2019.
https://doi.org/10.1007/978-3-319-98618-0_10

2 Literature Review

Delay at roundabout is commonly estimated based on capacity and volume-to-capacity ratio. The delay estimation methods described in several guidelines or manuals are reviewed below.

AUSTROADS
In AUSTROADS [3], roundabout delay is estimated based on Troubeck's steady-state theory and was firstly reported in the Australian Road Research Board Special Report No. 45 [4]. The initial queue is not considered in this formula. Equation (1) shows the average delay d_{st} (sec) at a roundabout entry described by steady-state delay model:

$$d_{st} = d_m + \frac{3600 \cdot k \cdot x}{c(1-x)} \tag{1}$$

where:

d_{st} - average delay [s],
d_m - minimum delay experienced by an entering vehicle in absence of queuing on the approach road [s],
k - delay parameter given by $k = \frac{d_m \cdot c}{3600}$,
x - volume-to-capacity ratio of the subject lane [-],
c - entry capacity [veh/h].

And the minimum delay d_m is estimated by Eq. (2) as shown below:

$$d_m = \frac{e^{\lambda(t_c - \tau)}}{\varphi q_{cir}} - t_c - \frac{1}{\lambda} + \frac{\lambda \tau^2 - 2\tau + 2\tau\varphi}{2(\lambda\tau + \varphi)} \tag{2}$$

where:

λ - a parameter in the exponential arrival headway distribution model,
τ - the minimum headway of circulating traffic,
φ - proportion of free vehicles,
t_c - critical gap which is calculated by t_f [s].

Parameters λ, φ, t_c and t_f are calculated by Eqs. (3)–(6):

$$\lambda = \frac{\varphi q_{cir}}{(1 - \tau q_{cir})} \tag{3}$$

$$\varphi = 0.75(1 - \tau q_{cir}) \tag{4}$$

$$t_c = (1.641 - 3.137 \cdot 10^{-4} q_{cir}) \cdot t_f \tag{5}$$

$$t_f = 2.819 - 3.94 \cdot 10^{-4} q_{cir} \tag{6}$$

In AUSTROADS [1], the entry capacity c is calculated by Eq. (7):

$$c = \frac{3600\varphi q_{cir} e^{-\lambda(t_c - \tau)}}{1 - e^{-\lambda t_f}} \tag{7}$$

SIDRA 4

In SIDRA 4 [5], the roundabout delay model was developed through incorporating the model described in AUSTROADS [3] and introducing a time-dependent delay formula [6, 7]. The model is shown in Eq. (8):

$$d = d_m + 900T \left[x - 1 + \sqrt{(x - 1)^2 + \frac{8kx}{cT}} \right] \tag{8}$$

where:

d_m - minimum delay from Eq. (2) [s],
T - flow period in hours, duration of the time interval during which the demand flow rate persists (1 h = 1),
x - volume-to-capacity ratio of the subject lane,
k - delay parameter from Eq. (1),
c - entry capacity from Eq. (7) [veh/h].

Highway Capacity Manual 6th Edition and HBS

In HCM 6th edition [8] and HBS [2], d_m can be calculated as $d_m = 3600/c$ by assuming delay parameter $k = 1$ (from Eq. (1)). Then, Eq. (8) turns to be Eq. (9) which is used to estimate roundabout delay in Highway Capacity Manual 6th edition. The formula is the same as the one shown in HCM 2010 [1]:

$$d = \frac{3600}{c} + 900T \left[x - 1 + \sqrt{(x - 1)^2 + \frac{\left(\frac{3600}{c}\right)x}{450T}} \right] + 5 \times \min[x, 1] \tag{9}$$

where:

d - average control delay [s/veh],
x - volume-to-capacity ratio of the subject lane [-],
c - capacity of the subject lane [veh/h],
T - time period [h] ($T = 0.25$ h for a 15-min analysis).

In this formula, the first two parts are same as that for stop-controlled intersections and the last part $5 \times \min[x, 1]$ is especially modified for yield control on the subject entry. As x increases, the probability of entry flow increased to approach to complete stop [1].

Through these estimation models, it is clearly known that capacity and volume-to-capacity ratio are significant variables for delay estimation. Thus, influencing factors such as pedestrians and geometric elements which have impact on capacity should also have impact on delay estimation. However, these influencing factors have not been considered in the existing formulas. Then, a simple way that inputting the capacity under the condition with pedestrians instead of the original one considering vehicles only is considered to reflect the pedestrian impact, but the existing delay estimation formulae may not be applicable because the queuing mechanism is changed, i.e. one or two additional stops can be caused by crossing pedestrians before exiting roundabout. Therefore, the purpose of this research is to estimate roundabout delay considering pedestrian impacts. In order to examine the applicability of the proposed theoretical formula, simulation study is firstly done, and the results of simulation are considered as a benchmark of this analysis.

3 Methodology

3.1 Simulation Study for Estimating Roundabout Entry Capacity

Simulation study is done including two conditions, vehicles only and with pedestrians. And the input parameters of the simulation study are calibrated based on the field data observed on roundabout in Japan.

Hypothesized Geometric Layout

Considering the generalization, a four-leg single lane roundabout with a diameter of 27 m is hypothesized in this analysis as shown in Fig. 1, which is the minimum requirement for the standard four-leg single lane roundabout in Japan. Splitter island is assumed to be present at each entry/exit.

Users of Roundabouts

In this analysis, in order to simplify the conditions, only passenger cars and pedestrians are assigned. Vehicle flows are assumed to be from each approach and pedestrian flows are assumed to exist at crosswalk of each entry/exit. Delay is observed for vehicles entering from Entry S.

Input Parameters

Speed

Referring to the explanations which were shown in the previous analysis of Kang and Nakamura [9–11], vehicle average speed in upstream of entry roadways are assumed to be 30 km/h with the uniform speed distribution. Speed reduction near yield line and in the circulatory roadway is realized by a function "Reduced Speed Areas" in VISSIM 5.40 and speed reduction on entry road is represented by deceleration as 2.0 m/s^2 under the condition of uniform deceleration. In the end, median of speed in the circulatory roadway are adjusted to 20 km/h for passenger cars. Furthermore, the minimum and maximum speeds in the circulatory roadway are set to be 18 km/h and 25 km/h dependent on field data. Pedestrian speed is assumed to be 1 m/s based on the author's previous research.

Parameter Settings for Gap Acceptance Behavior

Vehicle to Vehicle

Three parameters are included in the gap acceptance theory; critical gap t_c, follow-up time t_f and minimum headway τ. Based on the field data, values of t_c, t_f and τ are expected to be calculated to 4.5 s, 3.2 s and 2.2 s [11]. The function "Priority Rule" is selected to reflect critical gap t_c, and "Wiedemann 74 model" is used to reflect the other two parameters. The parameter "minimum gap time" in the "Priority Rule" model is calibrated to reflect t_c. In the "Wiedemann 74 model", there are three parameters, "average standstill distance (ax)", "additive part of desired safety distance (dx)" and "multiple part of desired safety distance (dx_mult)" [9, 10].

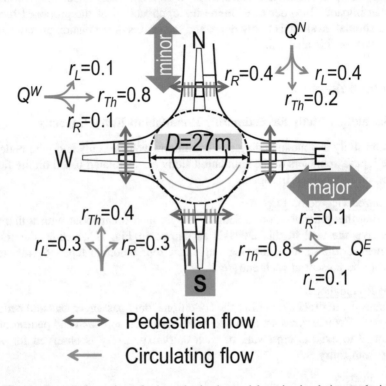

$$Q^N$$

$r_L=0.1$ $r_R=0.4$ $r_L=0.4$
Q^W $r_{Th}=0.8$ $r_{Th}=0.2$
$r_R=0.1$

$D=27m$

W   E

major

$r_{Th}=0.4$ $r_R=0.1$
$r_L=0.3$ $r_R=0.3$ $r_{Th}=0.8$ Q^E
$r_L=0.1$

S

⟷ Pedestrian flow

⟵ Circulating flow

Fig. 1. Basic information of a hypothesized roundabout in simulation analysis

Vehicle to Pedestrian
Based on the authors' previous research [11], entry capacity is influenced by the presence of splitter island, pedestrian approaching side and far-side pedestrian recognition rate. Thus, in this analysis, the simulation environment is assumed to be (1) with splitter island at each entry/exit; (2) pedestrians from both sides with identical volume and (3) far-side pedestrians yield rate = 1 (100% priority on crosswalk, vehicle will stop and wait a pedestrian completing the crossing on crosswalk). Under these conditions, critical gap of pedestrians was calibrated to be 6.20 s.

As the difference between target value and output value is lower than 10%, the calibration is considered as satisfactory.

Simulation Scenarios

Major-Minor Ratio and Turning Ratio

As shown in Fig. 1, in this analysis, vehicles are assigned from all approaches. Entries E and W are assumed to be the major approach and Entry N is assumed to be the minor approach. The ratio of major to minor traffic is assumed to be 8:2. Traffic volume and turning ratio of Entry E is assumed to be the same as Entry W, and the turning ratio of two entries is assumed to be $R_{LT}:R_{Th}:R_{RT}$ = 1:8:1 (LT: left turn, Th: through and RT: right turn). While, the turning ratios of Entry N and S are assumed to be $R_{LT}:R_{Th}:R_{RT}$ = 4:2:4 and 3:4:3, respectively.

Circulating Flow

For vehicle flow, based on the assumed major-minor ratio and turning ratios mentioned above, circulating flow can be arranged. And in this analysis, circulating flow is assigned to be 4 levels, 200, 400, 600 and 800 veh/h.

Pedestrian Flow

Pedestrians are assumed to come from both sides with the same volume, and the total pedestrian volume of one crosswalk is assigned to be 4 levels, i.e. 0, 50, 150 and 200 ped/h. Moreover, all pedestrians are given 100% priority (pedestrians will directly cross the crosswalk without waiting and until finishing crossing no vehicle will cross the crosswalk).

v/c Ratio

Regardless of the conditions with/without pedestrians, v/c ratio of Entry S is set to be 5 levels, 0.1, 0.3, 0.5, 0.7 and 0.9. The value is set for measuring input entry flow at Entry S. the actual v/c output is an approximate value.

Entry Flow at Entry S

Entry flow of Entry S is input by $c*(v/c)$. However, the values of entry capacity c under the condition only vehicle and with pedestrians are estimated separately as below.

(1) Under the condition of vehicles only

The setting of simulation study in this analysis is same as the ones of the previous research [9]. In the previous research, the validation results of simulation showed that the entry capacity of simulation study followed the ones which are estimated based on German formula as shown in Eq. (10). Thus, the entry capacity c under the condition of vehicles only in this analysis is calculated by this equation:

$$c = \frac{3600}{t_f}\left(1 - \tau\frac{q_{cir}}{3600}\right)\exp\left[-\frac{q_{cir}}{3600}\left(t_c - \frac{t_f}{2} - \tau\right)\right] \tag{10}$$

(2) Under the condition with pedestrians

The entry capacity under the condition with pedestrians is estimated by a regression model which was proposed by authors in their previous research [12]. The proposed model showed better fit to observed data than the results which are estimated by using the pedestrian adjustment factor indicated in HCM 2010 [1]. The results of the regression model are briefly introduced as follows.

The proposed regression model used the results from simulation and the shape of the model is shown in Eq. (11):

$$c_{ped} = A \exp(-Bq_{cir}) - Cq_{cir} \exp(-Bq_{cir}) \tag{11}$$

A, B and C are parameters which are determined by the identified factors x_i ($i = 1$, $2, \ldots, n$) and shown by Eqs. (12)–(14):

$$A = f(x_1, x_2, \cdots, x_n) \tag{12}$$

$$B = g(x_1, x_2, \cdots, x_n) \tag{13}$$

$$C = h(x_1, x_2, \cdots, x_n) \tag{14}$$

Four influencing factors are considered in this model, pedestrian demand at Entry S (ped/h), pedestrian demand at Exit W (ped/h), far-side pedestrian directional ratio r_{far} ([0, 1]) and far-side pedestrian yield rate FPYR ([0, 1]) are represented by x_1, x_2, x_3 and x_4.

After testing several types of regressions, e.g. linear, exponential and logistic regressions, it was shown that the linear regression gave the best fit result. According to this, linear regression functions of A, B and C are assumed and shown in Eqs. (15)–(17):

$$A = a_0 + a_1 x_1 + a_2 x_2 + a_3 x_3 + a_4 x_4 \tag{15}$$

$$B = b_0 + b_1 x_1 + b_2 x_2 + b_3 x_3 + b_4 x_4 \tag{16}$$

$$C = c_0 + c_1 x_1 + c_2 x_2 + c_3 x_3 + c_4 x_4 \tag{17}$$

where:
a_i, b_i, c_i ($i = 0, 1, \ldots, 4$) - coefficients.

The model is proposed separately under the condition with/without splitter island and the parameter estimation results of the model are referred to the results shown in the reference [12].

Based on the assumed simulation environment in this analysis, the parameters far-side pedestrian directional ratio r_{far} and far-side pedestrian yield rate FPYR are given the value of 0.5 and 1, respectively, and the corresponding results of the regression model can be obtained.

For every combination of input conditions, the VISSIM model was run for 10 times with a random number seed and for each run 1 h 15 min simulation time is assigned with a 15 min warm-up time. Data from the first 15 min of warm-up time were excluded from the results. Performance statistics were measured at a 15 min interval. In this analysis, the first 15 min of 1 h is used for observing delay, and delay is calculated by the travel time difference between the subject vehicle and the free flow vehicle. The measured delay (sec) was averaged based on 10 simulation runs.

4 Results and Discussion

4.1 Delay Considering Vehicles Only

Figure 2 shows the plots of delay results under the condition of vehicles only. It is found that average control delay increases when increasing circulating flow and also v/c ratio. The results are obviously reasonable since vehicles will spend more time to enter and exit roundabout when more vehicles exist in circulatory and entry roadways.

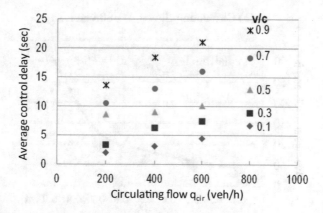

Fig. 2. Delay results from simulation study under the condition of vehicles only

Figure 3 shows the comparison results of simulation study (Sim results) and the ones which are estimated through applying the formula in HCM 6th edition (from Eq. (9)). The input parameter T of Eq. (8) is equal to 0.25 as the observation duration is 15 min in simulation study. The results under the condition that circulating flow is on the level of 200 veh/h and 800 veh/h are shown in Fig. 3(a), (b), (c) and (d), respectively. It is found that regardless of circulating flow levels, when v/c is less than 0.7, simulation results are slightly higher than those from HCM formula, but when v/c is higher than 0.7, the results from HCM 6th edition is extremely higher than those of the simulation study. It can be concluded that when v/c on the low and medium levels, results from simulation follow the ones that are estimated by HCM formula.

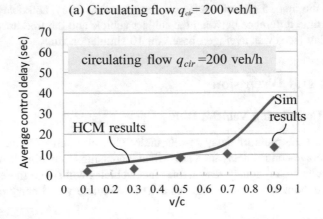

(a) Circulating flow q_{cir}= 200 veh/h

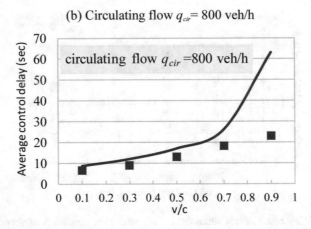

(b) Circulating flow q_{cir}= 800 veh/h

Fig. 3. Comparison results of delay between simulation results and HCM formula

4.2 Delay with Pedestrians

As mentioned previously, through applying the entry capacity regression model of Eq. (11) in delay estimation model (Eq. (9)), delay is calculated by Eq. (18) as shown below:

$$d_{ped} = \frac{3600}{c_{ped}} + 900T \left[x_{ped} - 1 + \sqrt{ \left(x_{ped} - 1 \right)^2 + \frac{\left(\frac{3600}{c_{ped}} \right) x_{ped}}{450T} } \right] + 5 \times \min \left[x_{ped}, 1 \right]$$

$$(18)$$

where:

d_{ped} - average control delay considering pedestrian [s/veh],
x_{ped} - volume-to-capacity ratio of the subject lane,

c_{ped} - capacity of the subject lane considering pedestrians from Eq. (11) [veh/h],
T - time period (T = 0.25 h for a 15-min analysis) [h].

Figure 4 shows the comparison results of delay estimation considering pedestrians from simulation study and theoretical formula. The delay results under the condition with pedestrians are shown in Fig. 4, and the results of circulating flow on the levels of 200 veh/h and 800 veh/h are shown in Fig. 4(a) and (b) as examples, respectively.

Firstly, through the simulation results, it is found that at each level of circulating flow, when increasing v/c ratio or pedestrian volume, also delay increases. The changing tendency is similar as the ones under the condition of vehicles only as shown in Fig. 2.

(a) Circulating flow q_{cir}= 200 veh/h

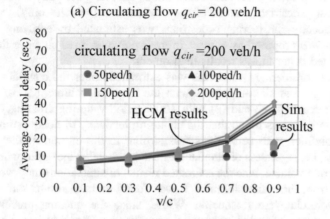

(b) Circulating flow q_{cir}= 800 veh/h

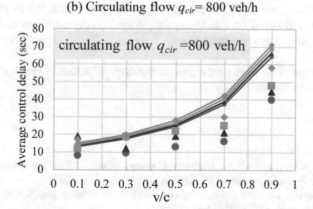

Fig. 4. Delay results considering pedestrians through applying simulation

Then, estimation results of delay by applying theoretical formula have shown the same changing tendency comparing to the results from simulation study, i.e. at any level of circulating flow, delay increases when increasing v/c ratio or pedestrian flow.

Then, the changing range of delay results from theoretical formula is narrower than the ones of simulation study regardless of v/c ratio. Moreover, similar to the results of considering vehicles only, at any level of circulating flow, when v/c is higher than 0.7, the difference between the results from simulation and theoretical formula becomes larger. Thus, it can be concluded that the theoretical formula can reasonably reflect the pedestrian impact on delay estimation.

5   Conclusion and Future Works

This paper estimated delay at roundabout considering pedestrian impact. Microscopic simulation VISSIM was applied in order to show a visual result. Besides applying microscopic simulation, a theoretical formula by inputting the entry capacity under the condition with pedestrians instead of the original one (vehicles only) was proposed. The entry capacity considering pedestrians was estimated by applying a regression model which was proposed by authors in their previous research since the regression model provided better fitness results than others.

The delay considering pedestrians was estimated under the conditions (1) pedestrians are assigned at all entries/exits with the identical volume, (2) from both sides with the identical volume and (3) with 100% priority during crossing. Through the simulation study, it was found that the changing tendency of delay estimation results under the condition with pedestrians was similar to the one under the condition of vehicles only, i.e. regardless of circulating flow level, delay increased when increasing v/c or pedestrian volume. And the estimated results of simulation study were compared to the ones from theoretical formula. It was found that the results through applying theoretical formula were reasonable. While, since the queuing process under the condition only vehicles is different from the condition with pedestrians, i.e. one stop by circulating flow and one or two additional stops by pedestrians, the theoretical model should be improved based on this viewpoint. Moreover, comparison to the field data should also be considered in future work.

Acknowledgements. This work was supported by JSPS KAKENHI Grant-in-Aid for Young Scientists (B), Grant Number 17K14743.

References

1. Transportation Research Board: Highway Capacity Manual 2010. Transportation Research Board of the National Academy of Science, Washington (2010)
2. FGSV: Handbuch Füer die Bemessung von Straßen. Forschungsgesellschaft Für Strassen - und Verkehrswesen. FGSV, Cologne (2005)
3. Austroads: Guide to Traffic Engineering Practice Part 6 - Roundabouts. Austroads, Sydney (1993)
4. Troutbeck, R.J.: Evaluating the Performance of a Roundabout. Special Report SR45. Australian Road Research Board, Australia (1989)
5. SIDRA for Roundabouts Manual (2011). http://www.sidrasolutions.com/Software/INTERSECTION/SIDRA_For_Roundabouts

6. Akcelik, R.: Implementing roundabout and other unsignalized intersection analysis methods in SIDRA. Working Paper WD TE91/002. Australian Road Research Board, Australia (1991)
7. Akçelik, R., Troutbeck, R.: Implementation of the Australian roundabout analysis method in SIDRA.   http://www.sidrasolutions.com/Cms_Data/Contents/SIDRA/Folders/Resources/Articles/Articles/~contents/QXF24HTELRWTL3XU/1991_Akcelik_Troutbeck_Implemnt_Aus_Rou_analysis-inSIDRA.pdf
8. Transportation Research Board: Highway Capacity Manual, 6th edn. Transportation Research Board, National Research Council, Washington (2016)
9. Kang, N., Nakamura, H.: An analysis of heavy vehicle impact on roundabout entry capacity in Japan. Transp. Res. Proc. **15**, 308–318 (2016)
10. Kang, N., Nakamura, H.: An analysis of characteristics of heavy vehicle behavior at roundabouts in Japan. Transp. Res. Proc. **25**, 1485–1493 (2017)
11. Kang, N., Nakamura, H., Asano, M.: Estimation of roundabout entry capacity under the impact of pedestrians by applying microscopic simulation. Transp. Res. Rec. **2641**, 113–120 (2014)
12. Kang, N., Kanbe, N., Nakamura, H., Odaka, S.: Development and validation of a roundabout entry capacity model considering pedestrians under Japanese conditions. Asian Transp. Stud. **4**(2), 350–365 (2016)

Models of Critical Gaps and Follow-up Headways for Turbo Roundabouts

Elżbieta Macioszek[(✉)]

Faculty of Transport, Silesian University of Technology, Katowice, Poland
elzbieta.macioszek@polsl.pl

Abstract. Turbo roundabouts are the new generation of circular intersections. They offer numerous advantages, the main of which include relatively high flow capacity and high traffic safety, the latter being only slightly lower than in one-lane roundabouts, still incomparably higher than that of multi-lane intersections. The first ever turbo roundabout was designed in the second half of the 1990s in the Netherlands by L. Fortuijna. Ever since, intersections of this type have been designed in many countries all around the world, and one of those where turbo roundabouts are created is Poland. This article describes models used to estimate psychotechnical parameters, i.e. critical gaps and follow-up headways for turbo roundabouts and for different traffic organisation scenarios with regard to the area where a minor road enters the main carriageway. The models in question have been designed with reference to results of surveys conducted at turbo roundabouts in Poland.

Keywords: Turbo roundabouts · Psychotechnical parameters at roundabouts Critical gap models · Follow-up headway models · Road traffic engineering Transport

1 Introduction

Turbo roundabouts are the new generation of circular intersections. The first ever turbo roundabout was designed in the Netherlands by L. Fortuijna in 1996 [1–3], and as stated in paper [4], as many as 70 turbo roundabouts had already functioned in the Netherlands by the year 2000. Offering numerous advantages, intersections of this type have been designed in many countries all around the world ever since. The group of countries where turbo roundabouts are created now also includes Poland. In 2018, there were about 150 such intersections in Poland. A small number of the newly built turbo roundabouts commissioned in Poland have been designed in line with specific requirements that turbo roundabouts must meet, as described in papers [2, 3, 5], However, a decided majority of the Polish turbo roundabouts currently in operation have emerged as a result of traffic organisation changes and adjustments to the geometry of formerly existing two-lane roundabouts. The foregoing causes that the circulatory roadway geometry often fails to conform with all the requirements that the circulatory roadway geometry in a typical turbo roundabout must meet. Furthermore, not all Polish turbo roundabouts feature traffic lane dividers protruding above the pavement level, and where this is the case, this function is merely performed by the P-2

© Springer Nature Switzerland AG 2019
E. Macioszek et al. (Eds.): Roundabouts as Safe and Modern Solutions in Transport
Networks and Systems, LNNS 52, pp. 124–134, 2019.
https://doi.org/10.1007/978-3-319-98618-0_11

type single solid line. The foregoing implies that one of the most fundamental design rules applicable to such intersections is neglected. From among all Polish roundabouts currently in operation, where a chosen traffic direction has been vested with the right of way, one can isolate a certain group of roundabouts characterised by mixed features of different roundabout types, both in terms of geometry and traffic organisation. All these aspects combined have led to a situation where, besides typical turbo roundabouts functioning in Poland, there are also junctions merely referred to as turbo roundabouts, while in fact they share very few characteristics with these intersections. Following a comprehensive survey of the circular intersections existing in Poland, it was possible to distinguish between two groups of Polish turbo roundabouts, and they are:

- turbo roundabouts with elevated traffic lane dividers protruding above the pavement,
- turbo roundabouts where the traffic lane divider function is performed by the P-2 type single solid line.

There are publications elaborating upon this subject which address the problems of traffic stream modelling from the macroscopic perspective (these models are characterised by the highest aggregation of traffic streams and the lowest accuracy, while for the sake of description of traffic stream behaviours, they rely on analogies to fluid mechanics) as well as from the mesoscopic perspective (characterised by high aggregation, low accuracy and analogies to gas kinematics used to describe behaviour of traffic streams), where turbo roundabouts are but one of transport network elements taken into consideration [6–16].

One may think of several typical types of traffic organisation at turbo roundabouts in the area where a minor road enters the main carriageway. This article describes models proposed to be used for estimation of some psychotechnical parameters, namely critical gaps and follow-up headways for turbo roundabouts on different traffic organisation scenarios with regard to the area where a minor road enters the main carriageway. The models in question have been designed with reference to results of surveys conducted at turbo roundabouts in Poland.

2 Characteristics of Turbo Roundabouts

A turbo roundabout is a multi-lane intersection featuring spiral circulatory roadway marking and traffic lanes separated for selected routes. Turbo roundabouts (similarly to the spiral ones) give privilege to selected traffic directions (only in one-lane and two-lane roundabouts, traffic participants have equal rights at all entry points). Depending on the number of traffic lanes at entries and exits, a roundabout may be configured in such a manner which precludes turning back at one of its directional routes. The main features of a turbo roundabout are as follows [17, 18]:

- more than one traffic lane functioning in the circulatory roadway,
- possibility to select the driving direction only at an entry (it is impossible to select or change the driving direction once in the circulatory roadway, since vehicle streams of the inner and the outer lane never intersect),

- no more than two traffic lanes in the roundabout's carriageway in the exit area where drivers are obliged to give way to the vehicles already using the main carriageway,
- impossibility to turn back at one of traffic directions (where the ring lacks sufficient widening).

Turbo roundabouts offer the following advantages [3, 4, 19, 20]:

- giving right of way to no more than two streams of vehicles moving in separated traffic lanes as they enter the intersection's main carriageway from one of its arms,
- reduced number of collision points compared to multi-lane roundabouts with a comparable number of traffic lanes and even compared to certain other types of intersections (Table 1).
- high traffic safety at turbo roundabouts owing to the relatively small number of collision points typical of this intersection type. Turbo roundabouts offer increased traffic safety compared to regular intersections (ca. 70% drop in the number of fatal accidents or accidents with hospitalised injured persons) as well as to two-lane roundabouts (overall number of traffic incidents reduced by ca. 50%). Turbo roundabouts are also characterised by higher traffic safety than intersections with traffic lights (ca. 50% drop in the number of fatal accidents or accidents with hospitalised injured persons). Nevertheless, turbo roundabouts are still ca. 20–40% less safe than one-lane roundabouts;
- relatively low driving speed at the roundabout (comparable to the vehicle driving speed typical of one-lane roundabouts) due to highly specific intersection geometry as well as elevated lane dividers (Fig. 1);
- high flow capacity of turbo roundabouts. The flow capacity of turbo roundabouts is higher than that of a one-lane roundabout, and in most cases, also a two-lane roundabout. In many cases, it is similar to or higher than the flow capacity of a signal-controlled intersection;
- lower time loss suffered by drivers compared to signal-controlled intersections;
- space and land demand for construction of a turbo roundabout (expressed in m^2) comparable to signal-controlled intersections, assuming that it is possible to enable parallel traffic of two heavy goods vehicles in any chosen direction at an intersection controlled by traffic lights. The costs involved in construction of a turbo roundabout are higher compared to a signal-controlled intersection, yet the related maintenance costs and social costs are lower.

Table 1. Number of collision points in selected types of intersections.

Number of intersection arms	Number of collision points					
	Regular intersections	Channelised intersections	Central island intersections	One-lane roundabouts	Two-lane roundabouts	Turbo roundabouts
3	9	9	–	6	16	7
4	32	32	36	8	22	10

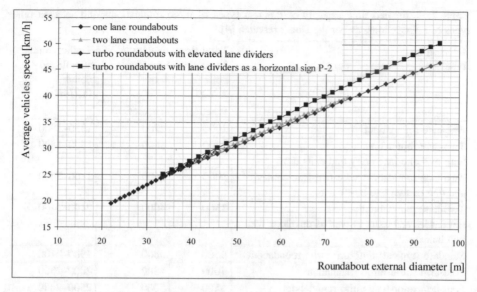

Fig. 1. Correlation between average driving speed of passenger cars using one-lane, two-lane as well as turbo roundabouts and external diameter (assuming the lane width in the main turbo roundabout carriageway of 4.5 m)

Turbo roundabouts may be classified as knee-type, star-type, standard, egg-type, spiral and rotor-type intersections. The respective names of individual turbo roundabout variations are related to the predominant vehicle stream direction at the intersection. The main factors decisive of the choice of the given turbo roundabout type under specific road and traffic conditions include the routing of the predominant traffic direction, vehicle stream volumes, average time losses involved in passing the intersection, terrain conditions (limitations) as well as investment costs.

Vehicular traffic at turbo roundabouts proceeds according to a principle of acceptance of headways in the main stream. On account of the fact that these intersections have only been in operation in Poland since recently, except for individual studies, there is no broad scale research concerning the flow capacity of these roundabouts in Poland. Table 2 provides a comparison of flow capacity values estimated for different types of turbo roundabouts functioning in the Netherlands with other types of roundabouts and regular intersections.

3 Models of Critical Gaps and Follow-up Headways for Turbo Roundabouts - Worldwide State of the Art

The psychotechnical parameters of vehicle drivers who approach intersections without traffic lights as well as roundabouts are typically considered to be critical gaps and follow-up headways. The critical gaps parameter for vehicle drivers at the roundabout entry (t_g) is commonly defined as the value of headway between vehicles in the main

Table 2. Estimated flow capacity of different types of turbo roundabouts and selected other types of intersections, as per the Dutch research [4].

Intersection type	Total flow capacity of all entries in the rush hour (± 10% of AADT)		Traffic volume at collision points PCU·h^{-1}]
	Practical	Theoretical	
One-lane roundabout	2000	2700	1350–1500
Two-lane roundabout with two one-lane entry and exits arms	2200	3600	1500–1800
Two-lane roundabout with two-lane entries and one or two one-lane exists	3000	3600	1800–2000
Two-lane roundabout (all entries and exits as well as envelope with two lanes)	3500	4000	2100–2400
Standard (typical) four-arm turbo roundabout	3500	3800	1900–2100
Four-arm spiral turbo roundabout	4000	4300	2000–2300
Four-arm rotor-type turbo roundabout (three-lane entries and two-lane exits)	4500	5000	2500–2800
Signal-controlled turbo roundabout (with 2 traffic lanes at entry for each route)	8500	11000	4200
Non-signal-controlled intersection with traffic lanes separated for the leftward-headed route	1500	1800	1100
Four-arm signal-controlled intersection (with 3 traffic lanes at each entry - 1 per each route)	3500	4000	3800
Four-arm signal-controlled intersection (with 6 traffic lanes at each entry - 2 per each route)	7500	8000	3800

vehicle stream at the roundabout, such that each headway whose value is equal or greater will be used for performing a manoeuvre of entering the roundabout lane by the respective driver from the subordinated entry (average in statistic terms), whereas each smaller headway (preventing a driver from performing the intended manoeuvre) cannot be used. In turn, the follow-up headway (tf) is defined as the time between the departure of the first (previous) vehicle from the roundabout entry and the departure of the next vehicle using the same major-street headway under the condition of queuing at the roundabout entry. If the distance between vehicles in a traffic stream in the roundabout circulatory carriageway allows for entry of further vehicles from the queue, they pass through the roundabout carriageway edge at follow-upheadway, one after another. On very low traffic volumes for a circulatory carriageway of a roundabout, the follow-up headway value is primarily decisive of the roundabout entry capacity.

So far, the overall body of literature of the subject has been rather scarce in studies addressing determination of psychotechnical parameters of vehicle drivers at turbo roundabout entries. The foregoing is mainly due to the fact that turbo roundabouts are among the newest road features. A summary of the models contemporarily used to characterise the psychotechnical parameters for turbo roundabouts has been provided in Table 3.

Table 3. Models characterising critical gaps and follow-up headways for turbo roundabouts.

Author(-s)	Traffic control at the entry	Values of critical gap [s]	Values of follow-up headway [s]
W. Brilon et al. [21] (Germany)	1	- Left lane: 4.5 - Right lane: 4.5	- Left lane: 2.5 - Right lane: 2.5
	2	- Entry lane: 4.5	- Entry lane: 2.5
	3	- Entry lane: 4.3	- Entry lane: 2.8
	4	- Left lane: 4.0 - Right lane: 4.5	- Left lane: 2.6 - Right lane: 2.7
L. Fortuijn [22, 23] (Netherlands)	1	- Left lane: 3.30 ± 0.31 (site 1) - Left lane: 3.54 ± 0.25 (site 2) - Left lane: 3.60 ± 0.28 (site 3) - Right lane: 3.67	- Left lane: 2.26 - Right lane: 2.13
	2	- Entry lane: 3.37 s	- Entry lane: 2.31–2.52
	3	–	- Entry lane: 1.80–2.45
	4	- Left lane: 3.08 ± 0.53 (site 1) - Left lane: 3.22 ± 0.48 (site 2) - Right lane: 2.79	- Left lane: 2.26 - Right lane: 2.13
M. Guereri, R. Mauro, R. Parla, T. Tollazi [24] (Italy, Slovenia. Research results represent Slovenian conditions)	1	- Left lane - 5.48 ± 2.22 - Right lane - 4.18 ± 2.70	- Left lane - 2.71 ± 1.70 - Right lane - 2.60 ± 1.49
	3	- Left lane - 4.41 ± 2.76 - Right lane - 4.03 ± 2.65	- Left lane - 2.53 ± 0.77 - Right lane - 2.56 ± 1.14
E. Macioszek [25] (Poland)	1	- Left lane: 3.21–4.27 - Right lane: 3.59–4.35	- Left lane: 3.23–3.83 - Right lane: 3.37–4.76
	2	- Entry lane: 3.07–3.85	- Entry lane: 4.35–4.61
	3	- Entry lane: 3.07–4.50	- Entry lane: 4.35–4.62
	4	- Left lane: 3.28–4.66 - Right lane: 3.59–4.35	- Left lane: 5.08–5.41 - Right lane: 4.37–4.76

where:

Individual roundabout traffic control scenarios for the entry zone have been defined as follows:

• 1: two lanes at the entry, two lanes at the roundabout, with one starting at the entry,

• 2: one traffic lane at the entry and two at the circular carriageway, with the possibility to enter only the outer traffic lane of the circular carriageway,

• 3: one traffic lane at the entry and two lanes at the circular carriageway, with possibility to enter both lanes of the main carriageway,

• 4: two lanes at the entry, two lanes at the roundabout.

4   Models of Critical Gaps and Follow-up Headways for Turbo Roundabouts in Poland

The models discussed in the article, used to determine numerical values of psychotechnical parameters that characterise vehicle drivers at turbo roundabout entries, were developed with reference to results of empirical studies conducted in Poland at 23 turbo roundabouts located in both developed as well as open space areas. Tests of traffic streams were performed in the years 2007–2013. The traffic observations covered roughly several hundred thousand vehicles in total, and the following road traffic characteristics were measured:

- traffic volume in main carriageways of roundabouts and at entries, distinguishing between individual traffic lanes in terms of vehicle stream load,
- headways in the main carriageway,
- gaps rejected and gaps accepted by drivers using individual traffic lanes at the entry, which subsequently provided grounds for estimation of threshold headway values,
- headway between vehicles entering the main carriageway after queuing in entry traffic lanes,
- structure of vehicles by type and direction,
- driving speed and travelling speed.

Table 4 provides general characteristics of the testing ground.

The models were developed for a group of turbo roundabouts with elevated traffic lane dividers protruding above the pavement. They were further broken down depending on the system of traffic organisation at the entry and in the entry lane, and as needed, also in the given traffic lane route. These models have made it possible to establish numerical values of both parameters in question, namely critical gaps (t_g) and follow-up headways (t_f), for turbo roundabouts with the first circulatory carriageway radius ranging between 10.0 and 30 m. The models have been graphically illustrated in Figs. 2, 3, 4 and 5.

Table 4. General characteristics of testing ground.

Conditions	Parameter	Range of parameter variability
Road	First radius of circulatory carriageway R_1 [m]	10.0–30.0
	Width of main carriageway traffic lane [m]	3.0–5.0
	Entry width (one-lane and two-lane arms) [m]	3.0–8.0
	Left-hand sight distance at entry [m]	18.0–42.0
Weather	–	Sunny weather, without precipitation
Traffic	Traffic volume at entry (Maximum traffic volume pertains to a three-lane entry. Maximum traffic volume at a two-lane entry came to 2300 [PCU·h^{-1}])	36–2800
	Traffic volume at roundabout carriageway in collision zone [PCU·h^{-1}]	47–2500
	Share of heavy vehicles in total vehicle stream [%]	0.4–25.3
	Share of one-track vehicles in total vehicle stream (motorcycles and bicycles) [%]	0.1–3.4

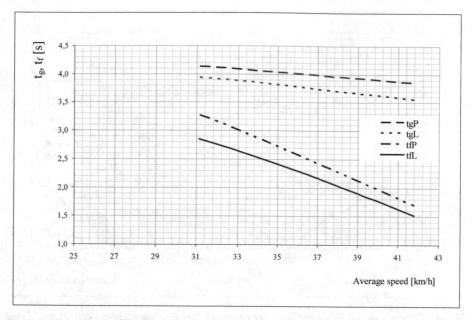

Fig. 2. Critical gap and follow-up headway functions for turbo roundabouts with elevated lane dividers in the case of two lanes at the entry and two lanes at the roundabout, with one starting at the entry (where: t_{gR} - critical gaps for right entry lane, t_{gL} - critical gaps for left entry lane, t_{fR} - follow-up headways for right entry lane, t_{fL} - follow-up headways for left entry lane)

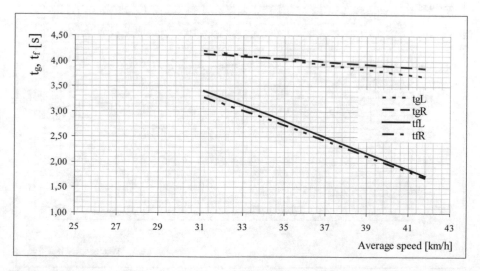

Fig. 3. Critical gap and follow-up headway functions for turbo roundabouts with elevated lane dividers in the case of two lanes at the entry and two lanes at the roundabout (where: t_{gR} - critical gaps for right entry lane, t_{gL} - critical gaps for left entry lane, t_{fR} - follow-up headways for right entry lane, t_{fL} - follow-up headways for left entry lane)

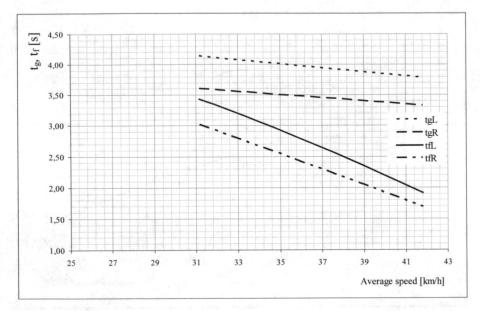

Fig. 4. Critical gap and follow-up headway functions for turbo roundabouts with elevated lane dividers in the case of one traffic lane at the entry and two lanes at the circulatory carriageway, where it is possible to enter both lanes of the main carriageway (where: t_{gR} - critical gaps for right entry lane, t_{gL} - critical gaps for left entry lane, t_{fR} - follow-up headways for right entry lane, t_{fL} - follow-up headways for left entry lane)

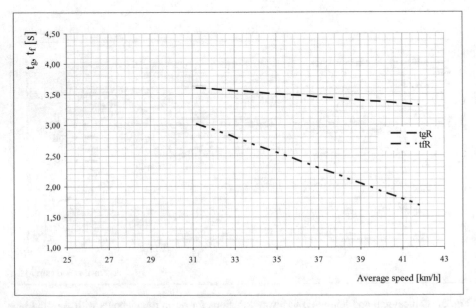

Fig. 5. Critical gap and follow-up headway functions for turbo roundabouts with elevated lane dividers in the case of one traffic lane at the entry and two at the circulatory carriageway, where it is possible to enter only the outer traffic lane of the circulatory carriageway (where: t_{gR} - critical gaps for right entry lane, t_{gL} - critical gaps for left entry lane, t_{fR} - follow-up headways for right entry lane, t_{fL} - follow-up headways for left entry lane)

5 Conclusions

The models discussed above have been positively verified under Polish conditions. However, on account of the fact that tests of parameters t_g and t_f were performed at turbo roundabouts in very early years after introducing this type of roundabout into Poland, it is recommended that further research should be conducted as turbo round-abouts continue to function in Poland. The preliminary period directly following the implementation of a new road organisation solution is typically the time when drivers become accustomed with it, learning how to properly use the given road feature. Meanwhile, their behaviour may be incorrect, slowed down or even inadequate to the actual potential of road infrastructure. As time passes and drivers grow in familiarity with the new solution as well as the principles which govern its use, their behaviour patterns become more representative. Hence the recommendation to continue this kind of research at turbo roundabouts in the incoming years.

References

1. Fortuijn, L.G.H., Harte, V.F.: Multi-Lane Roundabouts: Exploring New Models. CROW, Netherlands (1997)
2. Fortuijn, L.G.H.: Turbo-Roundabouts; Development and Experiences. http://www.bast.de/DE/Ingenieurbau/Publikationen/Veranstaltungen/B3-Downloads/turbo-kreisverkehre.pdf?_blob=publicationFile&v=1
3. Fortuijn, L.G.H.: Turbo roundabout: design principles and safety performance. J. Transp. Res. Board **2096**, 16–24 (2009)
4. Verweij, C.A., Boender, J., Coopmans, J.P.G., Drift, M.J.M., Fortuijn, L.G.H., Overkamp, D.P., Vliet, P., Wijk, W.: Roundabouts - Application and Design. A Practical Manual. DHV Group and Royal Haskoning. Ministry of Transport, Public Works and Water Management, Partners for Road, Holland (2009)
5. Giuffre, O., Guerrieri, M., Grana, A.: Turbo-Roundabout General Design Criteria and Functional Principles: Case Studies from Real World. http://www.4ishgd.webs.upv.es/index_archivos/75.pdf
6. Małecki, K., Wątróbski, J.: Cellular automaton to study the impact of changes in traffic rules in a roundabout: a preliminary approach. Appl. Sci. **7**(7), 742, 1–21 (2017)
7. Małecki, K.: The use of heterogeneous cellular automata to study the capacity of the roundabout. In: Rutkowski, L., Korytkowski, M., Scherer, R., Tadeusiewicz, R., Lotfi Zadeh, A., Zurada, J.M. (eds.) Artificial Intelligence and Soft Computing. LNAI, vol. 10246, pp. 308–317. Springer, Cham (2017)
8. Staniek, M.: Stereo vision method application to road inspection. Baltic J. Road Bridge Eng. **12**(1), 38–47 (2017)
9. Staniek, M.: Road pavement condition as a determinant of travelling comfort. In: Sierpiński, G. (ed.) Intelligent Transport Systems and Travel Behaviour. AISC, vol. 505, pp. 99–107. Springer, Switzerland (2017)
10. Staniek, M.: Moulding of travelling behaviour patterns entailing the condition of road infrastructure. In: Macioszek, E., Sierpiński, G. (eds.) Contemporary Challenges of Transport Systems and Traffic Engineering. LNNS, vol. 2, pp. 181–191. Springer, Switzerland (2017)

11. Sierpiński, G.: Technologically advanced and responsible travel planning assisted by GT planner. In: Macioszek, E., Sierpiński, G. (eds.) Contemporary Challenges of Transport Systems and Traffic Engineering. LNNS, vol. 2, pp. 65–77. Springer, Switzerland (2017)
12. Pypno, C.Z., Sierpiński, G.: Automated large capacity multi-story garage - concept and modeling of client service processes. Autom. Constr. **81C**, 422–433 (2017)
13. Villarreal Chávez, D.B., Kurek, A., Sierpiński, G., Jużyniec, J., Kielc, B.: Review and comparison of traffic calming solutions: Mexico city and Katowice. Sci. J. Silesian Univ. Technol. Ser. Transp. **96**, 185–195 (2017)
14. Turoń, K., Czech, P., Juzek, M.: The concept of walkable city as an alternative form of urban mobility. Sci. J. Silesian Univ. Technol. Ser. Transp. **95**, 223–230 (2017)
15. Turoń, K., Golba, D., Czech, P.: The analysis of progress CSR good practices areas in logistic companies based on reports "Responsible Business in Poland. Good Practices" in 2010–2014. Sci. J. Silesian Univ. Technol. Ser. Transp. **89**, 163–171 (2015)
16. Golba, D., Turoń, K., Czech, P.: Diversity as an opportunity and challenge of modern organizations in TSL area. Sci. J. Silesian Univ. Technol. Ser. Transp. **90**, 63–69 (2016)
17. Fortuijn, L.G.H.: Pedestrian and Bicycle-Friendly Roundabouts, Dilemma of Comfort and Safety. http://www.mnt.ee/atp/failid/SlowTrRoundb.pdf
18. Fortuijn, L.G.H., Carton, P.J.: Turbo Circuits: A Well - Tried Concept in a New Guise. http://www.pzh.nl/
19. Mauro, R., Branco, F.: Comparative analysis of compact multilane roundabouts and turbo roundabouts. J. Transp. Eng. - ASCE **136**(4), 284–296 (2010)
20. Szczuraszek, T., Macioszek, E.: Proportion of vehicles moving freely depending on traffic volume and proportion of trucks and buses. Baltic J. Road Bridge Eng. **8**(2), 133–141 (2013)
21. Brilon, W., Bondzio, L.: Experiences with Turbo-Roundabouts in Germany. In: 5th Rural Roads Design Meeting Copenhagen (2014). http://nmfv.dk/wp-content/uploads/2014/04/Experiences-with-Turbo-Roundabouts-in-Germany-Brilon-Bondzio-Weiser.pdf
22. Fortuijn, L.G.H.: Turbo Roundabout. Estimation Capacity. Transp. Res. Rec. **2130**, 83–92 (2009)
23. Fortuijn, L.G.H.: Turborotonde en Turboplein: Ontwerp, Capaciteit en Veiligheid. Trail Thesis Series 1, 1–367 (2013)
24. Guerrieri, M., Mauro, R., Parla, G., Tollazi, T.: Analysis of kinematic parameters and driver behaviour at turbo - roundabouts. J. Transp. Eng. Part A **144**(6), 1–12 (2018)
25. Macioszek, E.: Analysis of significance of differences between psychotechnical parameters for drivers at the entries to one-lane and turbo roundabouts in Poland. In: Sierpiński, G. (ed.) Intelligent Transport Systems and Travel Behaviour. AISC, vol. 505, pp. 149–161. Springer, Switzerland (2017)

Road Traffic Safety Analysis

Study on Roundabouts in Polish Conditions - Law, Safety Problems, Sanctions

Katarzyna Turoń[✉] and Piotr Czech

Faculty of Transport, Silesian University of Technology, Katowice, Poland
{katarzyna.turon, piotr.czech}@polsl.pl

Abstract. The present paper deals with the subject of the road infrastructure element, which is a roundabout intersection. The paper refers to the basic information and research conducted on the subject of roundabouts in international scientific publications. On their basis, a thematic niche concerning proper way to navigate the roundabouts and the interpretation of legal provision was noticed. Due to this fact, the authors analyzed the concept of 'roundabout' in terms of Polish legislation. In addition, the most frequent safety problems and mistakes made by drivers while driving through these types of intersections are described. In addition, the authors also referred to the sanctions that are applied to drivers in the event of improper use of roundabouts. The aim of the paper is to present the study on roundabouts in Polish conditions in the legal context, as well as to indicate the most common safety problems concerning the use of the roundabouts and sanctions related to improper drivers behavior when entering this type of intersection.

Keywords: Roundabout · Roundabout in polish legislation
Safety problems concerning driving through roundabouts
Intersections with circular motion

1 Introduction

Currently, in the era of increased burden of the road network, it is important to look for various types of solutions that are able to reduce traffic in urban transport systems [1, 2]. In addition, it is important that these solutions also have a positive impact on improving the level of road safety both with regards to the intersections and the entire section of the road [3, 4], be able to help improve traffic flows [5, 6] and become a practical solution for all road users including pedestrians and cyclists in accordance with the assumptions of sustainable development [7, 8]. One of such solutions is the use of roundabouts.

Roundabout are an element of road infrastructure that has been controversial for years. These controversies in Polish conditions are mainly related to legislative issues and the lack of appropriate regulations classifying the roundabout as a solution, which distinguishes itself from other types of intersections. Moreover, apart from legislative issues, the source of problems are also unregulated rules concerning driving through roundabouts as well as insufficient education regarding the methods of drivers behavior both when entering the intersection and while already driving at the intersection. Such

© Springer Nature Switzerland AG 2019
E. Macioszek et al. (Eds.): Roundabouts as Safe and Modern Solutions in Transport
Networks and Systems, LNNS 52, pp. 137–146, 2019.
https://doi.org/10.1007/978-3-319-98618-0_12

ambiguities become contentious issues, which pose problems especially during driving tests or in the case of unusual maneuvers performed by some drivers. Due to difficulties occurring in the interpretation of regulations and with regard to possible ways of driving at roundabouts, these intersections are regarded in the public opinion as not easy to handle with [9]. And as a result, these difficulties often result in the fines which are imposed on drivers because of the incorrect driving within this type of intersection.

Despite the hassles associated with the use of roundabouts, they constitute one of the safest types of intersections [1]. Therefore, the issue of this kind of intersection is reflected in many scientific studies conducted all over the world. Research topics of roundabouts cover the entire spectrum of aspects, which include: [9–16]:

- identification of factors that may affect the behavior of users when using this type of intersection,
- issues related to construction, appropriate geometry of intersection and appropriate signage due to the requirements of traffic engineering,
- proper adjustment of roundabouts to other elements of the infrastructure existing in the city,
- promotion of safety and proper behavior at such intersections due to mobility issues,
- traffic modeling on roundabouts and conducting simulation of mobility efficiency at such intersections, etc.

Despite the extensive research area both in the world and in Poland, considerations regarding legislative issues, principles defining the proper use of roundabouts and possible sanctions applied to drivers due to inadequate behavior at intersections, still remain a topic, which is not fully explored. Therefore the aim of the paper is to present the topic of roundabouts in Polish conditions in the legal context, as well as to indicate the most common safety problems occurring with regard to the use of roundabouts and sanctions related to improper driving on this type of intersection.

2   Roundabouts in the Context of Polish Legislation

Roundabout intersection is, according to the Polish language dictionary, "a circular square from which streets or roads radiate; also: crossing of two or more streets, formed in the shape of a square with the island in the middle and with the road surrounding it." [17]. An example of a roundabout is presented in Fig. 1.

The concept of roundabouts does not appear in Polish road traffic regulations. Moreover, there is no exact definition describing the roundabout. In the Act of 20 June 1997 of The Law On Road Traffic, it is able to find only single mentions concerning driving at intersections with circular motion, including the following ones: „It is forbidden to overtake a motor vehicle driving on the roadway at an intersection, except for intersections with circular motion or on which traffic is routed." (Art. 24.7.3) [18]. Legal regulations, which to a greater extent refer to intersections with circular motion, are the Regulation of the Minister of Infrastructure and Internal Affairs and Administration regarding road signs and signals - Dz.U.2002.170.1393 - Regulation of the

Fig. 1. An example of roundabout in city Myszków, Poland

Minister of Infrastructure and Internal Affairs and Administration of July 31, 2002 on road signs and signals [19].

The Act contains mentions of road signs placed before intersections with circular motion. In accordance with §5 on road signs within intersections [19]:

- the "intersection with circular motion" A-8 sign warns of an intersection where the traffic takes place around the island or square in the direction indicated by the sign,
- the "circular motion" C-12 sign indicates that at the intersection the traffic takes place around the island or square in the direction indicated by the sign.

In addition, it follows from the above mentioned regulation that the C-12 sign appearing together with the A-7 sign indicates the right of way of the driver at the intersection with regard to the driver who enters the intersection." [19]. This means that vehicles that already are present at the intersection have the right of way over vehicles that intend to enter the intersection. This rule also applies to rail vehicles (they are usually trams) [20].

According to the interpretation of the current regulations on moving at intersections and so-called "right hand rule" if no A-7 yield sign is placed before the intersection, then drivers entering the intersection are to yield to the drivers, which are already present at the intersection [20]. Summarizing the intersection where there is no sign A-7 is intersection with circular motion (not a roundabout). Examples of the application of the C-12 sign at intersections are presented in Figs. 2 and 3.

Fig. 2. An example of C-12 sign with A-7 sign on roundabout in Myszków, Poland

Fig. 3. An example of C-12 sign without sign A-7 on intersection in Sosnowiec, Poland

3 Safety Problems Regarding the Use of Roundabouts

Due to the lack of formulation of specific legal provisions referring to the roundabouts and intersections with circular motion, many drivers encounter problems with the correct use of such intersections. This is due to the fact that in order to thoroughly understand the rules to be followed, it is worth knowing the interpretation of the provisions of the Road Traffic Law with reference to classical intersections in order to be able to refer it to the roundabouts cases. The most common safety problems of drivers regarding the use of roundabouts are the following ones [21–24]:

- issues related to the right of way at the intersection,
- change of lane and rules concerning occupying lanes,
- entering and leaving the roundabout,
- turning back at the roundabout,
- using direction indicators in the situation of entering the roundabout, leaving the roundabout and changing the lane,
- correct behavior on roundabouts with traffic lights,
- correct behavior on turbo roundabouts and mini roundabouts.

According to the information previously provided, the right of way at the roundabout depends on the information indicated by the sign placed before intersection with circular motion. If there is A-7 "yield" sign accompanied with the C-12 "circular motion" sign, before the roundabout, then drivers entering the roundabout are obliged to yield the right of way to road traffic users, which are already present at the intersection. However, if there is only the C-12 sign located before the roundabout, the priority is given to users entering the intersection, and users, which are already present at the roundabout are obliged to allow the former ones to join the traffic [20].

Entering the roundabout, the change of the lane at the intersection and the rules of occupying the appropriate lane are carried out according to the rules in force when using the standard intersection [20] This means that if a given roundabout has [20]:

- one entry lane and one lane at the roundabout - the user of the road traffic occupies it after making sure that he/she is able to enter the intersection if the A7 „yield" sign is located before the roundabout or he/she occupies a lane in accordance with the rules regulating the right of way if this type of sign is not placed before the roundabout,
- one entry lane and two or three lanes at the roundabout - then the driver before entering the roundabout with the A7 "yield" sign has the obligation to yield to all users located on each lane of the roundabout - not only on the lane he/she intends to join,
- two entry lanes and two lanes at the roundabout - then the user entering the intersection with the A7 "yield" sign, continues on the lane he has chosen (right or left); the exception is the situation in which the driver is alone before the intersection, then he/she can take any lane of his/her choice,
- two entry lanes and three lanes at the roundabout - then the user entering the roundabout with the A7 "yield" sign, which occupies the left lane can take the left

lane at the roundabout, and the user entering the intersection with the right lane can take either the middle or the right lane,

- three entry lanes and three lanes at the roundabout - it is then important that the user of each lane, when entering the roundabout with the A7 "yield" sign, should take his lane respectively, continuing to drive or acting in a similar way to the situation in subsect. 4.

Turning round at the roundabout is one of the more problematic maneuvers due to the need to use direction indicators. Up to now, it was considered that the left direction indicator should be turned when turning round [20]. Such practice, however, led to misleading other road users. Furthermore, in accordance with the judgment the court in Lublin [25] and the judgment of the court in Gliwice [26] it is not necessary to use the left direction indicator when performing the turning round maneuver [25, 26]. In addition, it should be mentioned that it is not necessary to turn on the direction indicators when entering the roundabout. However, this is a procedure required when leaving the roundabout using a specific exit [20]. In this case, the maneuver should be signaled using the right direction indicator [20].

The issue of using direction indicators is also problematic when changing lanes. Then, following the common rules for lane change, one should indicate the desire to change the lane by turning on the appropriate direction indicator and yield to the vehicle, which is already on the lane that the user wants to enter [19]. Remember that when leaving the roundabout using the left lane you should signal the maneuver using the direction indicator and yield to the other road users [20].

A separate problematic group are situations when tracks enabling passage of rail vehicles run through intersection. In this case a few key rules should be remembered [20–24]:

- at the roundabout which is not equipped with traffic lights, the tram which enters the intersection is obliged to yield to vehicles that are already at the intersection,
- the tram at the roundabout has right of way over other vehicles; it means that the tram on entering the roundabout, which has no traffic lights, has right of way over the other vehicles at the roundabout regardless of the direction of its driving,
- if the tram is located at the roundabout equipped with traffic lights, then the traffic lights decide on the priority of crossing the intersection. If the tram is given a green light, then it has priority over other users.

The example of roundabout with the tram was presented in the Fig. 4.

One of the other problems, that drivers may face are problems with the so-called turbo roundabouts. The turbo roundabout is a kind of multi-lane intersection with circular motion, which has a spiral signage of the circulatory lanes and lanes designated for specific relations [27]. When using this type of roundabout, it should be remembered that when entering the roundabout, the user of the road traffic entering the intersection must determine the direction of travel by selecting the appropriate lane, so it is important to carefully read the information presented in the road signs, which are located before the roundabout; failure to comply with a specific directives indicated by the information sign may result in choosing the wrong lane and in consequence in lack of possibility of turning back at the crossroads as it is possible at the traditional

Fig. 4. An example of the roundabout with tram, Warsaw Poland

roundabout [20]. In addition, it should be remembered that in the case of a turbo roundabout, the yielding of right of way will take place only during the entry of road traffic users at the intersection [20]. Therefore, this kind of intersection is usually considered collision-free [27, 28].

The next group of problematic intersections also includes the mini roundabouts. Due to their construction, i.e. a small island that is not highly elevated [1], some road users consider this type of island as a land elevation through which one can drive straight ahead [20, 21]. It is worth remembering that performing such a maneuver may lead to failing to yield right of way and, as a result, to a collision. Therefore, it is important to remember that the rules of passing through mini roundabouts do not differ from the rules applicable in the case of classic large roundabout intersections or intersections with circular motion.

According to the Traffic Fine Schedule, incorrect behavior of road users at intersections with circular motion may result in imposing fines on them [29]. Selected penalties for incorrect behavior of road users at intersections, including intersections with circular motion, are presented in Table 1.

Table 1. Penalties for incorrect behavior of drivers at intersections, including roundabout intersections (Source: author's own collaboration based on [29])

Type of law violation	The amount of the penalty
Failure to comply with the C-12 "circular motion" traffic sign	250 PLN/5 penalty points
Inadequate vehicle positioning on the road before turning	150 PLN
Failing to yield the right of way to a rail vehicle	250 PLN/6 penalty points
Failing to yield the right of way at the intersection marked with signs that specify the right of way	do 500 PLN/6 penalty points
Failing to yield the right of way to the driver who changes the occupied lane	250 PLN/5 penalty points
Violation of the entry ban on the intersection, if there is no place in its area or behind it to continue driving	300 PLN/ 2 penalty points

4 Summary

To sum up, the paper presents the basic approach to the problem of intersections with circular motion according to Polish legislation. The examples of errors made by drivers and ambiguities as for their use of direction indicators, lane changes or the right approach to the right of way issue, indicate the necessity of unification and consolidation of the present regulations, which should be transformed into the general rules concerning the use of roundabout. Failing that roundabout users would be subject to high sanctions, which they are often exposed to due to their own ignorance, which results from inadequate education. It is worth mentioning that the set of rules concerning driving on intersections with circular motion should be an important element of education for both future candidates for drivers and the present ones, primarily due to the improvement of road safety.

References

1. Macioszek, E.: The comparative analysis of selected technical elements applied to traffic calming on intersections. Sci. J. Silesian Univ. Technol. Ser. Transp. **70**, 55–62 (2011)
2. Sierpiński, G.: Model of incentives for changes of the modal split of traffic towards electric personal cars. In: Mikulski, J. (ed.) Transport Systems Telematics 2014. Telematics - Support for Transport. CCIS, vol. 471, pp. 450–460 (2014)
3. Macioszek, E., Lach, D.: Analysis of the results of general traffic measurements in West Pomeranian Voivodeship over the years 2005-2015. Sci. J. Silesian Univ. Technol. Ser. Transp. **97**, 93–104 (2017)
4. Macioszek, E., Czerniakowski, M.: Safety-related changes introduced on T. Kościuszki and Królowej Jadwigi streets in dąbrowa górnicza between 2006 and 2015. Sci. J. Silesian Univ. Technol. Ser. Transp. **96**, 95–104 (2017)
5. Alkhaledi, K.: Evaluating the operational and environmental benefits of a smart roundabout. S. Afr. J. Ind. Eng. **26**(2), 191–202 (2015)

6. Pypno, C.Z., Sierpiński, G.: Automated large capacity multi-story garage-concept and modeling of client service processes. Autom. Constr. 81C, 422–433 (2017)

7. Turoń, K., Czech, P., Juzek, M.: The concept of walkable city as an alternative form of urban mobility. Sci. J. Silesian Univ. Technol. Ser. Transp. 95, 223–230 (2017)

8. Turoń, K., Golba, D., Czech, P.: The analysis of progress CSR good practices areas in logistic companies based on reports "responsible business in poland. good practices" in 2010–2014. Sci. J. Silesian Univ. Technol. Ser. Transp. 89, 163–171 (2015)

9. Distefano, N., Leonardi, S., Pulvirenti, G.: Factors with the greatest influence on drivers' judgment of roundabouts safety. An analysis based on web survey in Italy. https://doi.org/10.1016/j.iatssr.2018.04.002

10. Anjana, S., Anjaneyulu, M.V.L.R.: Development of safety performance measures for urban roundabouts in India. J. Transp. Eng. 141(1), 401–406 (2015)

11. Staniek, M.: Detection of cracks in asphalt pavement during road inspection processes. Sci. J. Silesian Uni. Technol. Ser. Transp. 96, 175–184 (2017)

12. Okraszewska, R., Nosal, K., Sierpiński, G.: The role of the Polish universities in shaping a new mobility culture - assumptions, conditions, experience. case study of Gdansk university of technology, Cracow university of technology and Silesian university of technology. In: Proceedings of ICERI2014 Conference, pp. 2971–2979. ICERI Press, Seville (2014)

13. Staniek, M.: Moulding of travelling behaviour patterns entailing the condition of road infrastructure. In: Macioszek, E., Sierpiński, G. (eds.) Contemporary Challenges of Transport Systems and Traffic Engineering. LNNS, vol. 2, pp. 181–191. Springer, Cham (2017)

14. Dabbour, E., Al Awadhi, M., Aljarah, M., Mansoura, M., Haider, M.: Evaluating safety effectiveness of roundabouts in Abu Dhabi. International Association of Traffic and Safety Sciences Research. https://doi.org/10.1016/j.iatssr.2018.04.003

15. Staniek, M.: Stereo vision method application to road inspection. Baltic J. Road Bridge Eng. 12(1), 38–47 (2017)

16. Flannery, A.: Geometric design and safety aspects of roundabouts. Transp. Res. Rec. 1751, 76–81 (2001)

17. Roundabout definition in Polish language dictionary. https://sjp.pwn.pl/szukaj/rondo.html

18. The Chancellery of the Seym of the Republic of Poland: Act of 20 June 1997 of the The Law On Road Traffic. The Chancellery of the Seym of the Republic of Poland, Warsaw (1997)

19. The Chancellery of the Seym of the Republic of Poland: Regulation of the Minister of Infrastructure and Internal Affairs and Administration Regarding Road Signs and Signals. Dz.U.2002.170.1393. The Chancellery of the Seym of the Republic of Poland Warsaw (2002)

20. Drexler, Z.: Traffic Regulations with an Illustrated Commentary. Grupa IMAGE, Warsaw (2011)

21. Sulowski, S.: The art of driving by roundabout. Motor 19, 34–37 (2017)

22. Sulowski, S.: Roundabouts - be careful on trams. Motor 38, 42–44 (2017)

23. Auto World Service: How to Correctly Drive on Roundabout. http://www.auto-swiat.pl/prawo/jak-poprawnie-jezdzic-po-rondzie/q2yb6s

24. Sulowski, S.: Use of direction indicators at the roundabout. Motor 13–14, 39–50 (2018)

25. II SA/GI 888/16. Decision of the Voivodship Administrative Court in Gliwice. http://orzeczenia.nsa.gov.pl/doc/23F480BB10

26. III SA/Lu 326/17. Judgment of the Provincial Administrative Court in Lublin. http://orzeczenia.nsa.gov.pl/doc/216D3D01A6

27. Macioszek, E.: Stan Bezpieczeństwa Ruchu Drogowego Na Rondach Turbinowych w Polsce. Prace Naukowe Politechniki Warszawskiej **96**, 295–307 (2013)
28. Macioszek, E.: First and last mile delivery - problems and issues. In: Sierpiński, G. (ed.) Advanced Solutions of Transport Systems for Growing Mobility. AISC, vol. 631, pp. 147–154. Springer, Switzerland (2018)
29. Traffic Fine Schedule. https://taryfikatormandatow.pl

Conceptual Multi-criteria Analysis of Road Traffic Safety Improvement for a Two-Lane Roundabout

Tomasz Szczuraszek[1](✉) and Paweł Szczuraszek[2]

[1] Faculty of Construction, Architecture and Environmental Engineering,
University of Technology and Life Sciences, Bydgoszcz, Poland
zikwb@utp.edu.pl
[2] Road Engineering Office in Bydgoszcz, Bydgoszcz, Poland
p.szczuraszek@bid-bydgoszcz.pl

Abstract. The aim of this study is to provide a detailed analysis of causes of road traffic low safety level on two-lane roundabouts, on the example of "Ofiar Katastrofy Smoleńskiej" roundabout situated in Olsztyn and an analysis of the most convenient alteration of the roundabout in order to improve its safety. The authors' own original method was used to analyze the choice possibilities of the roundabout rebuilding concept, which provides a multi-criteria assessment of comparable design solutions. This method takes into account four basic criteria: road traffic safety, traffic capacity, drivers' time loss and economic factors involved in the process of rebuilding, with an assumption of priority of the road traffic safety criterion. 6 rebuilding design variants were considered, whereas four of them offered designs of a turbo roundabout, and two 2 were crossroads with a central island and traffic light. Studies show that in terms of the accepted criteria the most beneficial were solutions based on the turbo roundabout scheme.

Keywords: Transport · Traffic engineering · Traffic safety

1 Introduction

It is commonly believed that roundabouts are the safest crossroads [1, 2]. Regular analyses of road traffic safety performed at the University of Science and Technology in Bydgoszcz for road networks in many towns in Poland deny this opinion in relation to two-lane roundabouts. These analyses show that typical two-lane roundabouts, pose relatively the highest threat to road traffic users, for a given level of traffic load, as compared to other types of crossroads. Certainly, crossroads with a central island and extended, multi-lane entrance roads, without traffic light, which are definitely leaders in terms of this kind of threat, should be excluded from this group. However, these crossroads are rare in Poland as design directives do not allow to use them without traffic light [3].

The goal of this study is to provide a detailed analysis of causes of road traffic low safety of two-lane roundabouts, on the example of "Ofiar Katastrofy Smoleńskiej" roundabout in Olsztyn and present a proposal of analysis to be used for selection of the

© Springer Nature Switzerland AG 2019
E. Macioszek et al. (Eds.): Roundabouts as Safe and Modern Solutions in Transport
Networks and Systems, LNNS 52, pp. 147–156, 2019.
https://doi.org/10.1007/978-3-319-98618-0_13

best rebuilding variant for this crossroads in order to improve road traffic safety in this place.

The considered crossroads is located in the central part of Olsztyn (town with population of 170 thousand inhabitants), in the suburbs. The crossroads performs an important function in the basic road network of this town. It also carries the transit traffic. Hence, it is significantly burdened. In the periods of the morning and afternoon rush hours these traffic volumes are app. 3 200 v/h.

The analyzed crossroad is a big two-lane roundabout with external diameter of 78 m, which originally had two, two-lane entrance roads and two exit roads. This roundabout, however, was supposed to be rebuilt for two main reasons:

- alteration of one of the one the crossroads streets from single carriageway to a dual carriageway, which involved preparing the crossroad for taking over additional transit traffic on national road no. 16,
- extension of one of the access roads in response to permanent traffic jams in rush hours.

In effect of rebuilding the crossroads was given shape of a typical two-lane roundabout with a lane for turns right separated on one of the entrance roads - see Fig. 3a. In result of these works a significant increase in capacity, from 4 400 v/h in a morning rush hour to 5 300 v/h in the afternoon rush hour, was achieved. In relation to anticipated traffic intensity, a reserve of capacity was reached, 830 v/h in the morning rush hour up to 1 600 v/h in the afternoon rush hour.

2 Assessment of Road Traffic Safety for the Analyzed Roundabout

Analyses of road collisions from the period of app. 2.5 years during the crossroads service revealed that it is characterized by a very low level of traffic safety. Annual average number of collisions was 80, including:

- 77.1% side collisions of vehicles,
- 16.1% rear collisions of vehicles,
- 3.4% skidding off the road because of excessive speed,
- 7% hitting a pedestrian.

Figure 1 shows a map of road collisions and Fig. 2 - a map of concentration of collisions within the crossroads.

More than half of collisions that have occurred in the roundabout were caused by drivers leaving the roundabout from the internal traffic lane who cut off drivers of vehicles going straight on the external lane of the roundabout. These maneuvers usually caused side collisions of vehicles (nearly 52% of all the collisions on the crossroads) or more rarely - rear collisions. Another most frequent cause of road adverse accidents ($\sim 10\%$) was violation of right of way upon entering the roundabout, mainly upon entering from the external lane of the entrance road. Drivers entering the roundabout from this lane have limited visibility of vehicles moving on the roundabout which stand on the internal lane of the entrance road. The remaining causes of road adverse

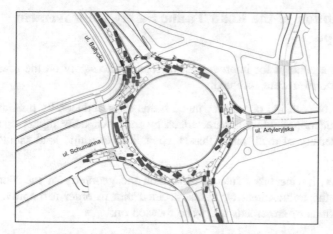

Fig. 1. Map of road collisions on "Ofiar Katastrofy Smoleńskiej" roundabout in w Olsztyn before alteration for a period of 2.5 years

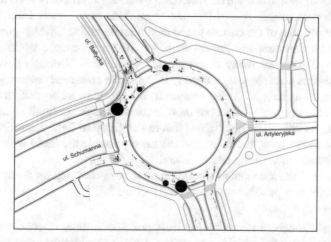

Fig. 2. Map of road collisions concentration on "Ofiar Katastrofy Smoleńskiej" roundabout in w Olsztyn before the alteration for a period of 2.5 years

accidents include cutting off each other while changing a traffic lane on the roundabout. The factors that contribute to these collisions are relatively high speeds of vehicles, geometry of the roundabout and inadequate traffic organization. Excessive speed was also the cause of vehicles skidding off and collisions with other road users (pedestrians, bikers).

Unfortunately, the extension of the roundabout enhanced its faults by introduction of two additional two-lane entrance and exit roads. By this, the level of collision risk on the roundabout increased up to ∼23%.

3 Description of the Road Traffic Safety Improvement Concepts

In search of a concept for improvement of road traffic safety on the roundabout, the following priorities were accepted:

- the level of collision occurrence threat needs to be reduced by minimum 50%,
- the roundabout should be characterized by traffic capacity higher than the antici-pated intensity of road traffic, that is, coefficient of traffic load should be smaller than 1.0.

Moreover, having taken into consideration the geometry of the roundabout and anticipating the traffic intensity, it was decided that in order to improve road traffic safety two kinds of crossroads had to be focused on:

- multi-lane turbo roundabout [2, 4–7],
- a crossroads with a central island and traffic light [3].

Both types of crossroads are characterized by a very high safety level and sufficient capacity. Whereas, the choice of the first crossroads type results from the fact that almost the whole area of the current roundabout can be utilized, which guarantees very low costs of its alteration and limited scope of road maintenance works. The second crossroads type provides many development possibilities. Through development of entrance roads it is possible to significantly increase the crossroads capacity. It occupies less area than a roundabout. It also enables in the future to incorporate it into a linear system of the area control which even more increases traffic continuity within the town road network. It needs to be highlighted that in the group of light controlled crossroads with multiple traffic lanes, the safest and the most passable are those with geometry of crossroads channeled with the central island.

Six variants of the roundabout rebuilding have been accepted for further analysis by considering different possible solutions [1–7] - Fig. 3.

Variant 'W1' - Fig. 3a

A turbine roundabout assuming the least possible number of changes to be intro-duced as compared to the initial state (variant 'W0') and with the lowest possible costs. The alteration involves changing traffic organization, exclusion of some surface parts from traffic, introduction of traffic separators for clear division of traffic channels and using paving stone for widening right turns in order to provide appropriate traffic channels.

Variant 'W2' - Fig. 3b

A turbine roundabout designed to minimize the number of changes in relation to the existing state and reduce the costs of implementation. The rebuilding covers a similar scope of work as variant I. It, however, differs by prioritization of Schuman - Arty-leryjska direction due to the highest traffic intensity and a planned run of national road no. 16.

Variant 'W3' - Fig. 3c

Turbine roundabout extended in relations to the initial state ('W0') by additional traffic lanes along the direction of Schumana and Artyleryjska streets - lanes added to

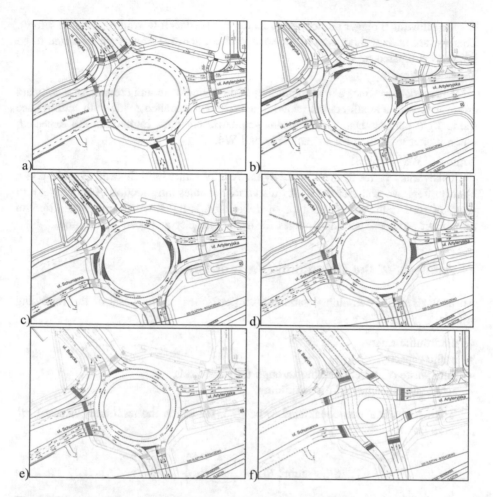

Fig. 3. Schemes of the considered rebuilding variants of "Ofiar Katastrofy Smoleńskiej" round about in Olsztyn: W0 (current state) - a, W1 - b, W2 - c, W3 - d, W4 - e, W5 and W6 - f

entrance roads of both streets as well as their prolongation within the area of the roundabout. Thanks to this the island has been given a shape of a typical turbine. The type of alteration is supposed to provide it with higher traffic capacity in the above mentioned direction. The scope of alteration, as compared to the previous variants, covers construction of the above mentioned additional traffic lanes and rebuilding of a two-lane app. 80 m. long section of Schuman street exit. Moreover, it involved exclusion of a separated turn right from Bałtycka street. Due to serious road resurface works the cost of rebuilding is significantly higher as compared to variants W1 and W2.

Variant 'W4' - Fig. 3d

This variant was designed with similar assumptions as the previous ones, with the only difference that the main goal is to provide the crossroads with higher capacity despite allowing a bigger number of collision points (thus lowering road traffic safety). This variant utilizes more area of the road surface, hence the costs are expected to be lower as compared to W3.

Variant 'W5' - Fig. 3e

This variant assumes a total rebuilding of the roundabout and creation of a compact crossroads with a small central island and expanded entrance roads with multi-phase traffic light control. Due to significant geometric changes, high costs are expected, much higher even than for variants W3 and W4.

Variant 'W6' - Fig. 3f

This variant varies from W5 only by the type of traffic light. Instead of multi-phase light it offers two-phase traffic light (this variant enables introduction of subphases). In terms of road traffic safety this type of traffic light is much worse than a multi-phase one, but it provides more advantages for traffic capacity and road users' time.

4   Analysis of the Rebuilding Variant Choice

An analysis of the rebuilding variants was performed on the basis of the following criteria:

- road traffic safety,
- traffic capacity,
- time waste of drivers passing through the crossroads,
- economic efficiency of the rebuilding.

The best variant was considered to be ('j'), for which the minimum value of SR_j measure is:

$$SR_j = \min\left\{ R_{j,1} + \frac{1}{3}\sum_{i=2}^{4} R_{j,i} \right\} \quad [-] \tag{1}$$

with simultaneous fulfillment of pre-established additional conditions:

- $PZ_j \leq 0.5PZ_{W0}$,
- $X_j < 1.0$.

where:

$R_{j,I}$ - position of the j-th variant in ranking 'i' (of a given criterion),
$R_{j,1}$ - position of the j-th variant in ranking no. 1, that is, "road traffic safety",
i - number of criterion according to the above mentioned order,
SR_j - value of the measure in the final ranking assessment 'j'- of this solution variant,
PZ_j - level of traffic intensity on a given j-th variant of crossroads [−],
X_j - level of traffic intensity on a given j-th variant of crossroads [−].

It needs to be emphasized that safety criterion SR_j is given the priority over the three remaining ones. The level of road collision occurrence risk, defined by the number of equivalent road accidents that happened on the crossroads in one year, was accepted to be a measure of road traffic safety.

A road collision with an average material loss was accepted to be an equivalent event. Estimation of a probable number of collisions for a given variant was performed on the basis of the method of potential collision points, in the following way:

$$PZ_j = \sum_{i=1}^{5} g_{ji} \times n_{ji} \quad [\text{equivalent road accidents}/\text{year}] \tag{2}$$

where:

n_{ji} - number of potential collision points of a given type 'i' on the j-th variant,

g_{ji} - weight of the i-th type of a potential collision point for the j-th variant, expressing a probable number of equivalent road accidents, which can generate such a point during one year.

Potential collision points include:

- intersection of two vehicle streams at a very small angle,
- intersection of two vehicle streams at an angle similar to right angle,
- intersection of vehicle streams with unprotected users of the roundabout (pedestrians and bikers),
- connection of two vehicle streams,
- disconnection of two vehicle streams.

Weights of particular kinds of potential collision points for particular types of crossroads were established on the basis of the authors' own tests performed for a few towns in Poland (crossroads with similar traffic intensity to the studied one). They also included data concerning the analyzed roundabout in Olsztyn, provided by the developed map of road collisions and the map of road collisions concentration.

Table 1 presents assessment of road traffic safety according to the above described method. The table also includes the state prior to the alteration - variant 'W_0' and after the alteration - variant 'W_0'.

Economic efficiency of particular variants for road traffic safety improvement was estimated by approximation using the following indicator E_j:

$$E_j = \frac{k_0 - k_j}{KB_j} \cdot \sum_{i=1}^{n} \frac{1}{\left(1 + \frac{r}{100}\right)} \ [-] \tag{3}$$

where:

k_o - yearly social costs connected with current operation of the crossroads (variant W_0),

k_j - yearly social costs connected with current operation of the crossroads (variant W_j),

KB_j - rebuilding costs according to variant 'j',

r - interest rate (discount) [%],

n - number of economic balance years - according to 'Blue Book' [8].

Table 1. Results of road traffic safety assessment for particular variants f the roundabout rebuilding.

Number of potential collision points

Variant no.	Intersections (small angle)	Intersections (big angle)	Intersections with unprotected users	Connection of streams	Disconnection of streams	Um
W_{00}	6.5	7	13	8	7.5	42
W_0	8	8	18	9	8	51
W_1	–	3	17	6	5	31
W_2	–	4	17	6	5	32
W_3	–	4	18	6	5	33
W_4	–	8	20	9	5	42
$W_{5(multi\ faz)}$	–	28	20	8	4	60
$W_{6(two\ faz)}$	–	28	20	8	4	60

Variant no.	Weights of potential collision points					Threat level [accidents/year]
	Intersections (small angle)	Intersections (big angle)	Intersections with unprotected users	Connection of streams	Disconnection of streams	
W_{00}	6.55	1.75	0.35	1.50	1.25	80.75
W_0	6.55	1.75	0.50	1.50	1.25	98.90
W_1	–	1.60	0.30	0.85	0.20	16.00
W_2	–	1.60	0.30	0.85	0.20	17.60
W_3	–	1.60	0.30	0.85	0.20	17.90
W_4	–	1.60	0.30	0.85	0.20	27.45
$W_{5(multi\ faz)}$	–	0.70	0.10	0.45	0.10	25.60
$W_{6(two\ faz)}$	–	1.24	0.15	0.50	0.15	42.32

The social costs included:

- costs of collisions,
- costs involved in waste of drivers' and car passengers' time,
- costs of the environment pollution,
- car running costs.

Cost of one equivalent road traffic event was assumed on the basis of data obtained from the Workshop of Economics of the Road and Bridge Research Institute in Warszawa. Whereas the remaining costs were assumed on the basis of data included in "Blue Book" [7].

The results of analyses regarding the choice of a rebuilding variant are contained in Table 2. They show that the most advantageous is variant W_1 - see Fig. 3, which provides the highest road traffic safety level and the best economic effects as compared

Table 2. Results of multi-criteria analysis of 'Ofiar Katastrofy Smoleńskiej' crossroads rebuilding variant.

No	Variant no.	Road traffic safety		Traffic capacity (year 2023)			Time losses		Economic effects		Sum of rankings SR$_i$ (final position)
		Level of threat [equv. of accidents/year]	Ranking	Traffic load degree	Traffic capacity margin [PCU/h]	Ranking	Per vehicle [s/PCU]	Ranking	Economic index. 'E'	Ranking	
1.	W_0	98.90	7	0.795	827	4	14.60	2	–	–	–
				0.667	1 611		5.40				
2.	W_1	16.00	1	0.795	894	5	15.80	3	456.50	1	4.0 (I)
				0.929	270		14.00				
3.	W_2	17.60	2	0.992	28	6	42.10	6	177.50	2	6.7 (II)
				0.930	265		15.70				
4.	W_3	17.90	3	0.995	18	7	34.60	5	20.20	4	8.3 (IV)
				0.927	272		14.70				
5.	W_4	27.45	5	0.730	1 284	2	7.10	1	46.50	3	7.0 (III)
				0.667	1 758		4.40				
6.	W_5	25.60	4	0.655 (0.596)	1030 (1477)	3	40.4 (34.7)	7	1.40	6	9.3 (V)
				0.728 (0.566)	591 (1764)		41.8 (36.0)				
7.	W_6	42.32	6	0.440	3 225	1	18.60	4	7.80	5	9.3 (VI)
				0.415	3 686		18.10				

to all the analyzed variants. However, in relation to the remaining variants, it is characterized by an average time loss of drivers and average traffic capacity.

5 Conclusions

- The research, performed on the basis of a selected crossroads, show that two-lane roundabouts with typical traffic organization are characterized by low level of safety which is also indicated in works [7, 9],
- the biggest disadvantage of two-lane roundabouts is a significant threat to drivers who leave the roundabout where vehicle streams cross at a small angle and where most collisions take place. Another serious disadvantage are two-lane entrance roads. In this case, vehicles standing on the left traffic lane of the entrance road block the view of drivers of vehicles standing on the right lane which causes frequent violations of right of way. Subsequently, there are more side collisions on the roundabout envelope in front of these entrance roads. It needs to be noted that a large diameter of a two-lane roundabout has also a negative influence on road traffic safety. It causes an increase in the vehicle speed and cutting off each other on the envelope, which leads to additional side collisions,
- analyses show that, turbine roundabouts provide high road traffic safety level and, due to appropriate geometry, enable wide selection of solutions to be adjusted to the capacity needs. As compared to traditional crossroads, equipped with traffic light, these crossroads also provide the possibility to reduce waste of drivers' time with simultaneous relatively similar traffic capacity and lower construction costs of (there is no need to install traffic light),

- the proposed method for choosing a rebuilding variant provides a multi-criteria assessment of comparable design solutions, with an assumption of road traffic safety priority. It seems to be an objective method and relatively simple to use,
- results of the authors' own research have shown that the places where two traffic streams cross pose the highest risk of collisions and they are characteristic of two-lane roundabouts. The risk is several times higher as compared to other potential collisions sites of crossroads.

References

1. Polish Association For Transportation Engineers: Design of Roundabouts - Experiences and New Trends. Research and Technical Papers of Polish Association for Transportation Engineers in Cracow. Polish Association For Transportation Engineers, Cracow (2010)
2. Macioszek, E.: The road safety at turbo roundabouts in Poland. Arch. Transp. **33**(1), 57–67 (2015)
3. General Directorate of Roads and Highways: Guidelines for the Design of Road Intersections. Part I, II. General Directorate of Roads and Highways, Warsaw (2001)
4. Verweij, C.A., Boender, J., Coopmans, J.P.G., Drift, M.J.M., Fortuijn, L.G.M., Overkamp, D. P., Vliet, P., Wijk, W.: Roundabouts - Application and Design. A Practical Manual. DMV Group and Royal Maskoning, Ministry of Transport, Public Works and Water Management, Partners for Roads, Holland (2009)
5. Grabowski, R.J.: Turbo - roundabouts as an alternative to standard roundabouts with the circular center Island. Roads Bridges **11**(3), 215–231 (2012)
6. Macioszek, E.: Analysis of the effect of congestion in the lanes at the inlet to the two-lane roundabout on traffic capacity of the inlet. In: Mikulski, J. (ed.) Activities of Transport Telematics. CCIS, vol. 395, pp. 97–104. Springer, Heidelberg (2013)
7. Bulla, L.A., Castro, W.: Analysis and Comparison Between Two-Lane Roundabout and Turbo Roundabout Base on Road Safety Audit Methodology and Microsimulation: A Case Study in Urban Area. http://onlinepubs.trb.org/onlinepubs/conferences/2011/RSS/2/Bulla,L. pdf
8. Jaspers: Blue Book - Joint Assistance to Support Projects in European Regions. Road Infrastructure. European Commission, Brussels (2015)
9. Tracz, M., Spławińska, M., Sakłak, W.: Bezpieczeństwo Ruchu na Rondach Dwupasowych. Transport Miejski i Regionalny **2**, 18–22 (2005)

Analysis of Pedestrian Behavior at Pedestrian Crossings with Public Transport Vehicles

Emilia Skupień[✉] and Mateusz Rydlewski

Faculty of Mechanical Engineering,
Wroclaw University of Science and Technology, Wroclaw, Poland
emilia.skupien@pwr.edu.pl,
mateusz.rydlewski@gmail.com

Abstract. The paper presents an analysis of pedestrian behavior at pedestrian crossing with public transport vehicles. That kind of pedestrian crossing are consider as with comparatively low number of vehicles passing through. The main goal of the analysis was to determine if a resignation of traffic lights on pedestrian crossings with public transport vehicle would decries the safety of pedestrians. The analysis was made based on the results of research conducted in the city of Wroclaw (Poland). The behaviors of pedestrian were divided into lawfully and illegal.

Keywords: Pedestrian behaviors · Pedestrian crossings
Crossing the public transport track

1 Introduction

Investigation of traffic engineering, taking into account pedestrian behaviour and safety, can be assigned to one of listed groups:

- analysis of digital images from measuring cameras, for analysing:
  - stopping a vehicle in front of pedestrian crossing,
  - retardation of a vehicle in front of pedestrians,
  - minimum distance between vehicle and pedestrian [1].
- measurements of luminosity at night, i.e. test of pedestrian visibility at the crossings,
- pedestrian surveys [2, 3],
- analysis of the consequences of accidents, for example resulting from the speed of vehicles [4],
- field research [5].

On the basis of only one of this four main research methods, data related to percentage of pedestrian passing through on a red light is possible. This is a field research. The example of such research are described in [5] where wide-angle cameras at the pedestrian crossings in Gdansk gave among others images of pedestrian entering the pedestrian crossing on the red light.

E. Macioszek et al. (Eds.): Roundabouts as Safe and Modern Solutions in Transport
Networks and Systems, LNNS 52, pp. 157–166, 2019.
https://doi.org/10.1007/978-3-319-98618-0_14

Most of the research related to digital camera images analysis refers to the driver-pedestrian relation and does not represent any significant facts of the behaviour of the pedestrian themselves [5].

The most significant disadvantage of analysis of the digital images is a problem with specifics of the examined places. To carry out the research and not being noticeable to traffic users, an infrastructure on which the cameras can be mounted is needed. It gives the possibility to perform the field research only at major intersections. This is related to the fact that on small intersections there is less infrastructure and its elements are smaller, which limits the points of locating the cameras.

As it is widely known, field research to be reliable should be conducted with an assurance of invisibility for the respondents. The responders might not act naturally and might make different decisions than in everyday situations. But in the matter of major intersections, a present of a digital equipment is harder to be sighted and even then it is received naturally.

Nevertheless a field research with the use of digital equipment is expensive due to the necessity of buying the equipment itself, its installation, costs of its software and programs to collect and analyse data. Research including observers (people who are in principle not visible to users of traffic) is less expensive and simpler to conduct, therefore it is considered more beneficial.

The field research including observers are instantly conducted in many countries [5]. However the surveys and analysing digital pictures and films are used more often to measure vehicle speed, priority giving, vehicle and pedestrian traffic flow and measure the time of crossing.

The European Sartre 4 Research Program is an element of the European Road Safety Observatory (ERSO). One of the question to the pedestrian was to assess a level of safety on the roads. With a choice of 3 descriptive options: (very/fairly, not much, not at all) Poland was 18th among 19 countries, with the following results:

- very/fairly 22%,
- not much 46%,
- not at all 32% [5].

At the forefront of European countries where pedestrians feel safe on the roads are such countries as Finland, Netherlands, Germany, Estonia and Sweden. Only Greece remains in the last place behind Poland. Due to such a low sense of security, conducting constant research is very important in particular when the results are published and gives the opportunity to react to disturbing signals.

In this paper a special attention was payed to the pedestrian crossing with public transport vehicles. This approach results from the fact that public transport vehicles are more predictable because:

- they stop on public transport stops (people seeing that vehicle in front of the stop, may assume that the vehicle will stop for exchange of passengers),
- relatively low acceleration compared to cars (especially near to stops),
- good visibility of approaching vehicles due to their dimensions,

- more often pedestrian has a small distance to overcome when passing the pedestrian crossing with public transport vehicles (2 lines) in opposite the pedestrian crossing with cars (e.g. 4 lines),
- public transport vehicles do not drive in columns like cars and there is a big probability, that despite the red signal for pedestrian, no vehicle arrives.

It all simplify crossing the pedestrian crossing even despite the red traffic light. The last reason for conducting this research is that the authors found a great lack of this type of research.

2   A Review of Polish Research Including Types of Pedestrian Behaviors

A field research including types of pedestrian behaviors of pedestrian crossings that were published, were conducted in Poland in 2009 [6], 2010 [7], 2015 [8], 2016 [9]. The results of research conducted in Wroclaw, excluding field research conducted on Reagan's roundabout, are shown in Table 1.

Table 1. Pedestrian behaviors on pedestrian crossing with public transport vehicles.

Year	Intersection	On green signal [%]	On red signal [%]	Sample size [Number of people]	Source
2009	Wita Stwosza - św. Katarzyny	43.0	57.0	35	[6]
2009	Wyszyńskiego - Szczytnicka	66.2	33.8	65	[6]
2009	Peronowa	42.0	58.0	112	[6]
2009	Piłsudskiego - Kołłątaja	54.3	45.7	99	[6]
2015	Sienkiewicza → Górnickiego	92.9	7.1	84	[8]
2015	Wyszyńskiego → Most Pokoju	45.6	54.4	64	[8]
2015	Sienkiewicza → Bema square	82.8	17.2	64	[8]
2016	Kościuszki square (1)	95.2	4.8	1329	[9]
2016	Kościuszki square (2)	89.1	10.9	1462	[9]
2016	Kościuszki square (3)	97.0	3.0	436	[9]
2016	Kościuszki square (4)	96.8	3.2	437	[9]

The results placed in Table 1 shows that most of pedestrian are passing pedestrian crossing in accordance with the low, but still there are some of crossings with alarmingly high rate of unlawful behavior. The research published in [5], developed by the public authority, on the basis of a study of behavior over 4200 pedestrians the following results were obtained:

- 8% of dangerous pedestrian behavior (crossing on a red traffic light), which is about 1% of all registered pedestrians. Such a low number results from the fact that 4200 pedestrians were studied in total and a behavior at pedestrian crossing with the traffic lights was observed on 1729 people (which constitutes 41% of the sample),

- among people passing on a red light 85.7% are people aged 20–60 (7.7% is <20 years old and 6.6% is <60 years old),
- these studies also show that more than 10% of pedestrians are passing through crossings with traffic lights without ensuring about the possibility of safe passage.

3 An Analysis of Pedestrian Behaviours at Pedestrian Crossing with Public Transport Vehicles on an Example of Reagan's Roundabout in Wroclaw

The research area is an important interchange node, where pedestrian safety is very important. The research shown in this paragraph were conducted on Reagan's roundabout which is a large interchange for public transport, with 4 platforms, and in the near area of a shopping mall, universities, dormitories and office buildings. At this roundabout, traffic lights are controlled by the ITS system. The research was carried out in 2010–2018. The authors focused on pedestrian crossing with public transport vehicles.

It also has to be noted that the measurement took place as a part of the students classes in the subject of the Basics of Traffic Engineering. Therefore, they were aimed primarily at acquainting the organization with measurements and traffic observation capabilities. The importance of a invisibility of people making measurements for traffic participants was not the first priority. It is obvious that this could affect the distortions of the measurement results by influencing the pedestrian behavior. Unfortunately over the years, it can be observed that such phenomenon is becoming rarer - nowadays, even when a police patrol is present, it is met to enter the pedestrian crossing on a signal prohibiting it.

The geometry of the intersection (roundabout), rebuilt in 2008, is shown on Fig. 1. Analysis and characteristics of individual crossings are shown below (Fig. 1):

1 - The only of the analyzed pedestrian crossings with public transport vehicles and two lines for cars.
2 - A tram crossing with two tram lines.
3 - A bus and tram crossing.
4 - A tram and bus crossing (4 lanes + 1 tram stop).
5–9 - A bus and tramway crossing with high intensity of public transport vehicles.

The number of attempts for particular pedestrian crossing and percentage share of pedestrians passing on various traffic signals are shown in the Table 2. The field research were conducted mostly in the afternoon (15–17).

Figure 1 shows a very high level of passing the pedestrian crossing on a red light on this type of intersections. Almost every fifth pedestrian breaks the law, most on the 4th crossing.

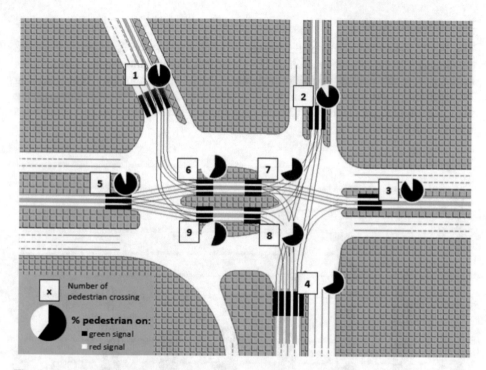

Fig. 1. Pedestrian behaviors on pedestrian crossing with public transport vehicles on Reagan's roundabout (results from 2010–2018)

Table 2. The number of attempts for pedestrian crossings on Reagan's roundabout percentage share of pedestrians passing on various traffic lights (results from 2010–2018).

Number of pedestrian crossing	Sample size	Number of measurements	Average number of people for measurement	Pedestrian on green signal	Pedestrian on red signal
	[Number of people]	[Items number]	[Number of people]	[%]	[%]
1	3893	25	155.7	97.3	2.7
2	684	4	171.0	91.5	8.5
3	191	2	95.5	90.6	9.4
4	2218	12	184.8	67.7	32.3
5	163	3	54.3	93.9	6.1
6	234	1	234.0	57.7	42.3
7	84	1	84.0	70.2	29.8
8	150	1	150.0	70	30
9	151	1	151.0	55	45
Total	7768	50	155.4	79.9	20.1

3.1   Data Analysis

The Fig. 2 shows the results of field research from a pedestrian crossing 1. It can be concluded that the implementation of ITS improved the operation of the traffic lights here ensuring the passage of public transport vehicles in a short time after receiving information from the detection sensor which results in the lack of pedestrians crossing the road, who in years before 2012 were used to stable traffic program and could have learned it by heart.

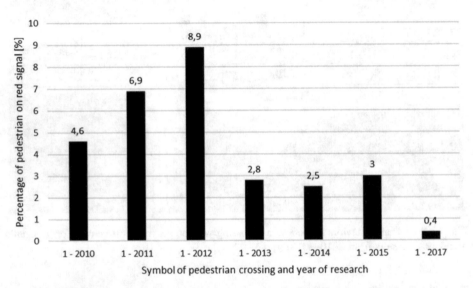

Fig. 2. The participation of pedestrians passing on red signal at the "1" pedestrian crossing at the Reagan's roundabout in 2010–2017

On Fig. 2 one can see that in 2017 a very low percentage of passing on red lights, most likely due to the creation of a new ground crossing leading from the shopping mall to the island and further to the other side of the square.

Figure 3 shows the results for 2 and 3 pedestrian crossing. The presented pedestrian crossing (number 2) has a collision with trams of 0 and 1 lines. The highest results for illegal behaviour for pedestrian crossing number 2 were noticed in 2016.

The pedestrian crossing number 2 leads directly from the University of Agriculture to the shopping mall and to the island with bus- and tram-stops. Hence the large share of pedestrian are probably students and young people. There is also a lack of traffic flow between traffic lights on following pedestrian crossings.

The relation on pedestrian crossing number 3 shown on Fig. 3 has a very few measurements, the passage is isolated from the middle part of the intersection. The intensity of traffic on this side of Grunwaldzki Square is relatively low compared to the other directions (no streams generating traffic).

Figure 4 shows the results from measuring pedestrian behaviors on pedestrian section 4 and 5. In the results from the crossing number 4 there is a lack of data for

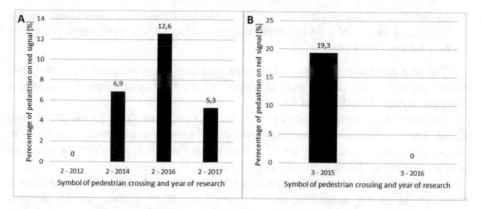

Fig. 3. The participation of pedestrians passing on red signal at the "2" (A) and "3" (B) pedestrian crossings at the Reagan's roundabout in 2012–2017

2012, 2013, 2014 and 2016 (there were no measurements). Nevertheless it can be seen that the passing on red traffic light tends to decrease. However the percentage of illegal behaviors is high - almost every 5 pedestrian passes on red. This may be due to the fact that the stop of tram line 1 is the only one that does not stop at the roundabout island.

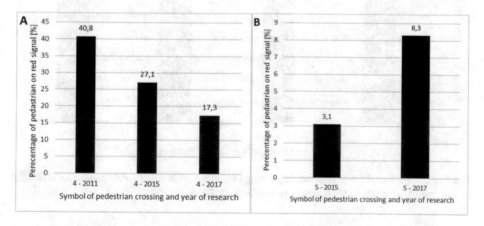

Fig. 4. The participation of pedestrians passing on red signal at the "4" (A) and "5" (B) pedestrian crossings at the Reagan's roundabout in 2011–2017

There is a high percentage of pedestrian crossing passage 4 on red light. It might be due to the fact that:

- there are only 3 bus lines passing there and 4 tram lines (trams are considered to be more predictable by pedestrians),
- it is wide (4 lanes), but one in each direction is used only by 1 public transport line, therefore this part of the road is usually used as an asylum during red light display,

- behind this crossing is a stop of tram line 1 and as it departures quite seldom, at the moment of its arrival a lot of people may be willing to risk crossing on red not to wait for the next tram.

The next pedestrian crossing 5 is shown on Fig. 4. A significant increase in the percentage of passing on red, may be caused by introducing an additional transition from the asylum directly to the island.

The other character of the pedestrian crossing are between stops, on the interchange island. The results from measurement can be influenced by a significant increase in pedestrian traffic at the roadway level (no need to use the underpass) since the launch of over ground routes leading directly to the island were established.

Common directions (routes) of vehicles passing from different platforms encourage pedestrian to change depending on the time of arrival of a given public transport vehicle. Results are shown on the Fig. 5.

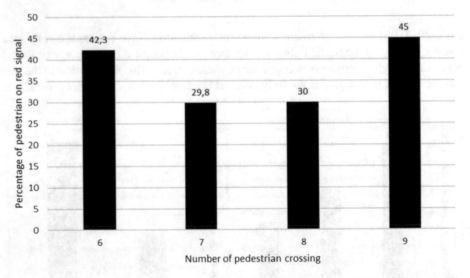

Fig. 5. The participation of pedestrians passing on red signal at the "6"–"9" pedestrian crossings on the interchange island at Reagan's roundabout in 2018

Definitely more pedestrians pass the pedestrian crossings on the red traffic light on the left due to the fact that on this side are two new ground transitions leading directly to the island. On the right side there is only one such passage.

Results are distressing - 4/10 pedestrian passes on red traffic light. The main reasons may be as follow:

- the arrival of the vehicle on which the pedestrian waits, to be on another platform,
- no worries to cross the narrow section of the road,
- very good visibility of public transport vehicles due to their dimensions,

- the specificity of public transport vehicles movement is small acceleration and low speed of both starting (when starting from the stop) as well as access to the stop where all vehicles are pausing to stop.

4 Conclusion

In this paper authors shown the problem with pedestrian crossing at public transport vehicles ways with traffic lights. The results of analysis show that pedestrian are less likely to follow the rules if they feel safe.

The most percentage of pedestrian who cross a pedestrian crossing on red signal is on the central island of roundabout. These values reach up to 45% - pedestrian crossing number 9 (Fig. 5). This is probably due to a very small section of the road to be overcome and the haste resulting from the direct neighborhood of public transport stops. For this reason it is very important to ensure pedestrian safety. The solution to this problem may be the installation of special type of signaling with sound transmitters which can alert people from the danger, such as vehicles of public transport.

On several pedestrian crossings there is a decreasing trend regarding pedestrian crossing on red signal. These are pedestrian crossings number 4 (Fig. 4A), 3 (Fig. 3B) and 1 (Fig. 2). In the case of pedestrian crossing number 1, since 2013 one can notice a decrease in the number of pedestrians, who break the traffic rules. The reason for this occurrence may be the system ITS which has been implemented at this roundabout in 2012.

The paper presents partly results of students work, therefore it is possible to suggest cooperation between the urban engineering departments of the cities concerned, together with universities and technicians conducting classes on subjects related to traffic engineering.

If it comes to interchange island - one can apply solutions consisting in the liquidation of traffic lights and the activation of an audible and visual warning in the event of approaching the vehicle following the example of cities such as Budapest, Vienna or even recently Gdansk [10]. In case of Gdansk, on some pedestrian crossings, it was decided to completely shut down the traffic lights [11]. This action is aimed at accelerating public transport vehicles and ensuring freedom of movement for pedestrians.

The use of this type of solution would have to be characterized by high reliability. E.g. adaptation of the action to traffic conditions and walking, too early warning will not make sense and too late can lead to unsafe situations.

At the 3 pedestrian crossing, one can modify the signaling program. It is a pedestrian crossing on which not in every cycle there is public transport vehicle which may encourage pedestrians to go on red light. In such situations, a phase that would not activate red for pedestrians should be used.

The most important is proper synchronization with neighboring transitions to allow the pedestrian crossing the road at one cycle, not with the need to stop and wait for a new one.

References

1. Olszewski, P., Czajewski, W., Dąbkowski, P., Szagała, P.: Badanie Zachowań Uczestników Ruchu Na Przejściach Dla Pieszych Na Podstawie Analizy Obrazu. Budownictwo i Architektura **13**(4), 177–184 (2014)
2. Hatfield, J., Fernandes, R., Job, R.F.S., Smith, K.: Misunderstanding of right-of-way rules at various pedestrian crossing types: observational study and survey. Accid. Anal. Prev. **39**(4), 833–842 (2007)
3. Sullman, M.J.M., Abigail, T., Stephens, A.N.: The road user behavior of school students in Belgium. Accid. Anal. Prev. **48**(9), 495–504 (2012)
4. Richards, D., Cuerden, R., Hill, J.: Pedestrians and Their Survivability at Different Impact Speeds. https://pdfs.semanticscholar.org/a01f/5d61153084bf36d18337b2e1f8f2f161e231. pdf
5. Ministerstwo Infrastruktury i Rozwoju: Metodologia Systematycznych Badań Zachowań Pieszych i Relacji Pieszy-Kierowca Wraz z Przeprowadzeniem Badań Pilotażowych - Etap 1. Założenia do Metodologii wraz z Badaniami Pilotażowymi. Ministerstwo Infrastruktury i Rozwoju, Warszawa (2015)
6. Skupień, E.: Wpływ Sposobu Funkcjonowania Sygnalizacji Świetlnej na Zachowania Pieszych na Przykładzie Placu Grunwaldzkiego we Wrocławiu. Praca Dyplomowa Magisterska. Politechnika Wrocławska, Wrocław (2009)
7. Molecki, B., Skupień, E.: Zachowania Pieszych w Obrębie Dużych Skrzyżowań z Sygnalizacją Świetlną na Przykładzie Placu Grunwaldzkiego we Wrocławiu. Transport Miejski i Regionalny **6**, 31–36 (2010)
8. Podobiński, J.: Propozycja Zmiany Funkcjonowania Sygnalizacji Świetlnej W Celu Poprawy Ruchu Komunikacji Zbiorowej i Pieszej w Obrębie Wybranego Skrzyżowania. Praca Dyplomowa Inżynierska. Politechnika Wrocławska, Wrocław (2015)
9. Rydlewski, M.: Propozycja Zmiany Funkcjonowania Sygnalizacji Świetlnej w Celu Poprawy Ruchu Komunikacji Zbiorowej i Pieszej w Obrębie Placu Kościuszki we Wrocławiu. Praca Dyplomowa Inżynierska, Politechnika Wrocławska, Wrocław (2016)
10. Baltic Journal. http://dziennikbaltycki.pl/motofakty/aktualnosci/a/tramwaje-pojada-szybciej-na-al-zwyciestwa-w-gdansku-zniknie-czesc-swiatel-sterujacych-ich-ruchem,12565350/
11. Website. https://www.trojmiasto.pl/wiadomosci/Zdejmuja-sygnalizatory-by-przyspieszyc-tramwaje-n96195.html

Analysis of Traffic Safety on Roundabouts in Kędzierzyn-Koźle (Poland)

Damian Lach[✉]

Faculty of Transport, Silesian University of Technology, Katowice, Poland
damian.lach@polsl.pl

Abstract. Roundabout as a circular intersection can be treated depending on the country as one intersection or as a set of connected intersections. The traffic on the roundabout is around the island or around this part. According to the definition, roundabouts at this type of intersections minimize the number of collision points is making this kind of crossing safer than other types of intersections. The article contains an analysis of traffic safety on roundabouts in the city of Kędzierzyn-Koźle, in Poland in 2008–2017.

Keywords: Roundabout · Traffic safety · Safety analysis · Traffic incidents

1 Introduction

Roundabouts are often the subject of scientific research, e.g. [1–5]. The roundabout, depending on the country of its occurrence, may be characterized as one element of the road infrastructure, i.e. the intersection or as a set of such elements, i.e. a set of intersections dependent on each other. In the case of right-hand traffic, the traffic at the roundabout is counter-clockwise. The purpose of the roundabouts is to strive for a minimum of interlacing traffic flows. An important element of any roundabout intersection is the minimization of collision points, which means that this type of infrastructure element is considered to be very safe. Another task of the roundabout is to reduce the speed of vehicle traffic at the intersection, which contributes to the increase of safety in its area. In accordance with the regulation of the Ministry of Infrastructure and the Ministry of Interior and Administration regarding road signs and signals [6] a traffic sign C-12 should be placed at every inlet of the roundabout. The sign informs that the given infrastructure element should be treated as one intersection and not as a set of intersections. In addition, the C-12 road sign is used in conjunction with the A-7 sign on public roads, so that the driver at the intersection has priority over the driver who wants to enter the intersection. Unquestionable, on the safety also has an impact a vertical and horizontal signs. Road safety is strongly dependent on the speed of vehicle. Therefore, on the inlets of the roundabouts, speed-reducing elements such as canalization islands or elements causing optical narrowing of the lane are often used. The article presents the data of the Voivodeship Police Headquarters in Opole regarding the number of road incidents and other factors such as weather conditions or the type structure of a vehicle on roundabouts in the city of Kędzierzyn-Koźle. The result of the analysis is also the indication of the roundabout, which is characterized by the best

© Springer Nature Switzerland AG 2019
E. Macioszek et al. (Eds.): Roundabouts as Safe and Modern Solutions in Transport
Networks and Systems, LNNS 52, pp. 167–175, 2019.
https://doi.org/10.1007/978-3-319-98618-0_15

traffic conditions in terms of safety as well as the one on which the largest number of road accidents took place.

2   Analysis of Road Traffic Safety on Roundabouts

2.1   Analysis Area

The analysis covers all the roundabouts that function within the city of Kędzierzyn-Koźle. This kind of intersections have gained popularity in the analyzed city in the last few years. At the turn of the 20th and 21st centuries, the first Milenijne Roundabout was established at the intersection of national road No. 40 with voievodeship roads No. 423 and 408. For several years it was the only intersection of this type in the city. The next roundabout, which was established in Kędzierzyn-Koźle, was the Grunwaldzkie Roundabout at the crossing point of the national road No. 40 with the poviat road No. 14350. Both roundabouts have been built on the route of the central section of the city bypass and are designed to improve the traffic of vehicles on given routes. Since 2012, along with the reconstruction of the road infrastructure, further roundabouts have been systematically established. The third was a turbine roundabout at the intersection of Kozielska street and the city bypass, i.e. national road No. 40. The fourth roundabout was created at the intersection of Kozielska and Gliwicka streets. The next roundabout was created at the intersection of Kozielska, Reja and Tartaczna streets, near the railway station. The last two roundabouts that were created in the city center. One is located at the intersection of Aleja Jana Pawła II, Grunwaldzka and Karol Miarki streets, and the other at the intersection of Jana Pawła II, 1 Maja and Miłej streets. At present, there are 7 roundabouts in total in Kędzierzyn-Koźle. Table 1 presents a list of all analyzed roundabouts located in the city of Kędzierzyn-Koźle.

Table 1. List of roundabouts located in Kędzierzyn-Koźle.

No.	Name	Location	Type of roundabout	Diameter of roundabout [m]
1.	Milenijne	Armii Krajowej, Wyspiańskiego, Gliwicka intersection	Semi two-lane	38
2.	Grunwaldzkie	Armii Krajowej, Grunwaldzka intersection	Single-lane	35
3.	Pileckiego	Armii Krajowej, Kozielska intersection	Turbo	~50
4.	Roundabout no. 1 (unnamed)	Gliwicka, Kozielska intersection	Single-lane	30
5.	Wojaczka	Kozielska, Reja, Tartaczna intersection	Single-lane	35
6.	Solidarności	Jana Pawła II, Karola Miarki, Grunwaldzka intersection	Single-lane	40
7.	Roundabout no. 2 (unnamed)	Jana Pawła II, 1 Maja, Miła intersection	Single-lane	25

The inlets at all mentioned roundabouts have the appropriate markings, i.e. the A-7 road sign and the C-12 road sign. Figure 1 shows all roundabouts on the background of the city of Kędzierzyn-Koźle.

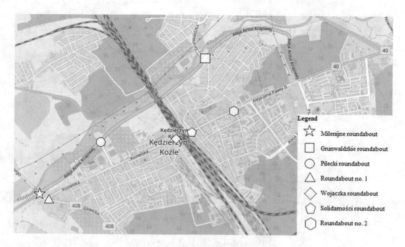

Fig. 1. Map of analyzed roundabouts in Kędzierzyn-Koźle, Poland (Source: [7])

Roundabout No. 1 and Rondo No. 2 are intersections that have not been given any name. Therefore, for the purpose of the analysis, the nomenclature was adopted in accordance with Table No. 1.

2.2 The Structure of Road Incidents

All data of road incidents that were presented in the article were made available by the Voivodeship Police Headquarters in Opole for the period from 2008 to 2017. This part of the article presents the number of road incidents that took place in the area of the roundabouts mentioned in Subsect. 2.1. Along with the number, the generic structure of vehicles participating in given road incidents was presented. Additional information, which is shown is the condition of weather conditions. Table 2 shows the number of road incidents on individual roundabouts. They were divided in the years in which they occurred.

According to Table 2, the majority of road incidents in the analyzed period occurred at the Milenijne Roundabout. The next roundabout with a large number of events relative to other roundabouts is Grunwaldzkie Roundabout. Figure 2 show the graph with data from Table 2.

Table 3 presents the generic structure of vehicles participating in road incidents.

Table 3 shows that passenger cars are the largest group of vehicles participating in road incidents at the analyzed roundabouts. Table 4 presents the weather conditions that prevailed during the occurrence of particular road incidents.

Table 2. List of traffic incidents on roundabouts in Kędzierzyn-Koźle.

Year	Roundabout	No. of traffic incidents
2008	Milenijne	6
	Grunwaldzkie	2
2009	Milenijne	4
	Grunwaldzkie	2
2010	Milenijne	13
	Grunwaldzkie	8
2011	Milenijne	6
	Grunwaldzkie	4
2012	Milenijne	11
	Grunwaldzkie	5
	Pileckiego	4
2013	Milenijne	13
	Grunwaldzkie	5
2014	Milenijne	9
	Grunwaldzkie	1
	Pileckiego	1
2015	Milenijne	12
	Grunwaldzkie	8
2016	Milenijne	11
	Grunwaldzkie	5
	Pileckiego	4
	Wojaczka	1
2017	Milenijne	5
	Grunwaldzkie	1
	Solidarności	1

Table 4 shows that the most traffic incidents took place during good weather conditions. This may indicate that there are no connections that the weather conditions have affected the traffic incidents that have occurred.

2.3 Safety Analysis

The main objective of the analysis was to indicate the factors that most often occur during road accidents and to find links between the given events and external factors.

Based on the analyzes carried out, it can be concluded that the most traffic incidents took place at the Milenijne Roundabout. The Milenijne Roundabout is a semi two-lane roundabout, from here it can be inferred that the drivers are not able to navigate this type of roundabouts. From 2010 to 2016, the number of road incidents on the Milenium Roundabout has doubled in relation to previous years. Based on the available publications, this can be related to the increasing traffic of individual vehicles, which in turn leads to more traffic incidents [8]. Considering the case of a roundabout, one should

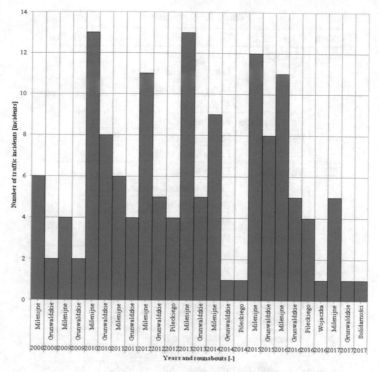

Fig. 2. Number of incidents on roundabouts in Kędzierzyn-Koźle, Poland

Table 3. Structure of vehicles of traffic incidents on roundabouts in Kędzierzyn-Koźle.

Year	Roundabout	Type of vehicle	No. of vehicles
2008	Milenijne	Passenger car	12
	Grunwaldzkie	Passenger car	4
2009	Milenijne	Passenger car	7
		Lorry with trailer	1
	Grunwaldzkie	Passenger car	4
2010	Milenijne	Passenger car	22
		Lorry without trailer	2
		Lorry with trailer	1
		Bus	1
	Grunwaldzkie	Passenger car	15
		Motocycle	1
2011	Milenijne	Passenger car	11
		Lorry without trailer	1
	Grunwaldzkie	Passenger car	8

(continued)

Table 3. (*continued*)

Year	Roundabout	Type of vehicle	No. of vehicles
2012	Milenijne	Passenger car	17
		Lorry without trailer	1
		Lorry with trailer	2
		Motocycle	2
	Grunwaldzkie	Passenger car	10
	Pileckiego	Passenger car	5
		Lorry with trailer	1
2013	Milenijne	Passenger car	22
		Lorry without trailer	1
		Lorry with trailer	1
		Motocycle	2
	Grunwaldzkie	Passenger car	9
		Motocycle	1
2014	Milenijne	Passenger car	17
		Lorry with trailer	1
	Grunwaldzkie	Passenger car	2
	Pileckiego	Passenger car	2
2015	Milenijne	Passenger car	23
		Lorry without trailer	1
	Grunwaldzkie	Passenger car	14
		Lorry with trailer	1
		Bus	1
2016	Milenijne	Passenger car	19
		Lorry with trailer	1
		Motocycle	2
	Grunwaldzkie	Passenger car	10
	Pileckiego	Passenger car	8
	Wojaczka	Passenger car	2
2017	Milenijne	Passenger car	12
	Grunwaldzkie	Passenger car	2
	Solidarności	Passenger car	2

Table 4. Weather conditions of traffic incidents on roundabouts in Kędzierzyn-Koźle.

Year	Roundabout	Type of weather	No. of traffic incidents
2008	Milenijne	Sunny	2
		Cloudy	4
	Grunwaldzkie	Sunny	2
2009	Milenijne	Sunny	2
		Rainy	1
	Grunwaldzkie	Sunny	2

(*continued*)

Table 4. (*continued*)

Year	Roundabout	Type of weather	No. of traffic incidents
2010	Milenijne	Sunny	5
		Rainy	2
		Cloudy	1
	Grunwaldzkie	Sunny	2
		Rainy	2
		Cloudy	1
2011	Milenijne	Sunny	2
		Cloudy	2
		Foggy	1
	Grunwaldzkie	Sunny	2
		Cloudy	2
2012	Milenijne	Sunny	4
		Cloudy	1
		Rainy	2
		Foggy	1
	Grunwaldzkie	Sunny	3
		Rainy	2
	Pileckiego	Sunny	2
		Cloudy	1
2013	Milenijne	Sunny	7
		Cloudy	3
	Grunwaldzkie	Sunny	2
		Cloudy	2
2014	Milenijne	Sunny	7
		Rainy	1
	Grunwaldzkie	Sunny	1
	Pileckiego	Sunny	1
2015	Milenijne	Sunny	7
		Cloudy	4
	Grunwaldzkie	Sunny	5
		Cloudy	1
2016	Milenijne	Sunny	2
		Cloudy	2
		Rainy	3
	Grunwaldzkie	Sunny	5
	Pileckiego	Sunny	3
	Wojaczka	Sunny	1
2017	Milenijne	Sunny	4
		Cloudy	1
		Rainy	1
	Grunwaldzkie	Sunny	1
	Solidarności	Rainy	1

also pay attention to the generic structure of vehicles participating in road incidents. In relation to other roundabouts on the Milenijne Roundabout, a large part of the vehicles are trucks with or without a trailer. Sources of the problem should be considered in the distribution of vehicle traffic to the network, because Armii Krajowej Avenue is a city bypass and a transit route for vehicles over 3.5 tons. The Milenijne Roundabout is also the only intersection of roads connecting the western, eastern and central parts of the city as well as the aforementioned transit route (national road No. 40).

Despite the fact that Grunwaldzkie Roundabout is also located on the city bypass, the number of road incidents is significantly lower. This is due to the smaller traffic prevailing on roads crossing the national road No. 40 in the place of Grunwaldzkie Roundabout. Roads are connected by districts whose population is relatively small.

The Pileckiego Roundabout, which is the third roundabout that was created in the analyzed city, is a turbo roundabout. This is the only representative of this type of roundabout in the city. At the indicated roundabout, the number of traffic incidents is small. This may result from the geometrical layout of a given roundabout and the way the traffic is driven by individual lanes. It should be noted that the given roundabout is also located on the national road No. 40 (the central section of the city bypass) where the main commercial and service zone generating significant traffic of individual vehicles as well as lorries. The dependence of a small number of traffic incidents and high traffic volume gives information confirming the beneficial effect of the use of turbo roundabouts which are separating traffic flows and increasing the capacity of a given intersection.

Roundabout No. 1, Wojaczek Roundabout, Solidarności Roundabout and Roundabout No. 2 are intersections that were rebuilt in 2016–2017. The scope of data of traffic incidents on the road infrastructure elements listed is therefore small. The only road incidents that were registered occurred at the Wojaczek Roundabout and the Solidarności Roundabout. Here you should look for connections with external factors such as the location on the main thoroughfare of the city (Kozielska street and Jana Pawła II Avenue) as well as the area in which the roundabouts were built. It is in fact the center of the city, in which the intensity of vehicle traffic is much greater than on the other mentioned roundabouts.

3 Conclusion

On the basis of the analysis the following conclusions were obtained:

- the majority of road incidents occurred on the Milenijne Roundabout, which is a semi two-lane roundabout,
- turbo roundabout is a significantly better solution than the semi two-lane roundabout. This trend is shown by the number of road incidents on Pileckiego Roundabout (turbo) and Milenijne Roundabout (semi two-lane),
- passenger cars account for over 90% of the total type structure of vehicles participating in road accidents in the analyzed period of time,

- about 64% of traffic incidents took place during good weather conditions, which means that weather conditions did not have a significant impact on the occurrence of particular events,
- in terms of roundabout type, the best solution is the turbo roundabout. Pileckiego Roundabout shows the smallest number of traffic incidents in terms of traffic intensity and the area in which it was built,
- the roundabouts located on Kozielska Street and Jana Pawła II Avenue should be thoroughly analyzed, as the short period of their functioning did not allow to collect data enabling proper assessment.

References

1. Macioszek, E.: The comparison of models for follow-up headway at roundabouts. In: Macioszek, E., Sierpiński, G. (eds.) Recent Advances in Traffic Engineering for Transport Networks and Systems. LNNS, vol. 21, pp. 16–26. Springer, Switzerland (2018)
2. Macioszek, E., Czerniakowski, M.: Road traffic safety-related changes introduced on T. Kościuszki and Królowej Jadwigi streets in Dąbrowa Górnicza between 2006 and 2015. Sci. J. Sil. Univ. Technol. Ser. Transp. **96**, 95–104 (2017)
3. Stuwe, B.: Capacity and safety of roundabouts - German results. In: Brilon, W. (ed.) Intersections Without Traffic Signals II, pp. 1–12. Springer, Berlin (1991)
4. Schoon, C., Minnen, J.: Safety of roundabouts in the Netherlands. Traffic Eng. Control **35**(3), 142–148 (1994)
5. Campbell, D., Jurisich, I., Dunn, R.: Improved Multi-lane Roundabout Designs for Urban Areas. New Zealand Transport Agency, New Zealand (2012)
6. Regulation of the Ministers of Infrastructure and of the Interior and Administration of July 31, 2002 on Road Signs and Signals. http://prawo.sejm.gov.pl/isap.nsf/download.xsp/WDU20021701393/O/D20021393.pdf
7. Open Street Map. http://www.openstreetmap.org
8. Macioszek, E., Lach, D.: Analysis of the results of general traffic measurements in the West Pomeranian Voivodeship from 2005 to 2015. Sci. J. Sil. Univ. Technol. Ser. Transp. **97**, 93–104 (2017)

Correction to: Making Compact Two-Lane Roundabouts Effective for Vulnerable Road Users: An Assessment of Transport-Related Externalities

Paulo Fernandes and Margarida Coelho

Correction to:
Chapter "Making Compact Two-Lane Roundabouts Effective for Vulnerable Road Users: An Assessment of Transport-Related Externalities"
in: E. Macioszek et al. (Eds.): *Roundabouts as Safe and Modern Solutions in Transport Networks and Systems*, **LNNS 52,**
https://doi.org/10.1007/978-3-319-98618-0_9

In the original version of the book, the following belated correction has been incorporated: The Acknowledgement section in Chapter 9 has been revised. The book and the chapter have been updated with the change.

The updated version of this chapter can be found at
https://doi.org/10.1007/978-3-319-98618-0_9

Author Index

© Springer Nature Switzerland AG 2019
E. Macioszek et al. (Eds.): Roundabouts as Safe and Modern Solutions in Transport
Networks and Systems, LNNS 52, p. 177, 2019.
https://doi.org/10.1007/978-3-319-98618-0

Printed in the United States
By Bookmasters